A Meteorite Killed My Cow

Meteorites are incredible objects, there's no doubt about it! But for most people, they seem far removed from their day-to-day lives. However, the truth is very different. Ordinary people encounter space rocks all the time, but are usually unaware of it. Until now. *A Meteorite Killed My Cow* will demonstrate that meteorites are the most important objects you will ever encounter. Without them, you wouldn't be here. And they are everywhere. You might be standing on one as you read this book.

Written in a clear and jargon-free style, *A Meteorite Killed My Cow* will introduce you to the fascinating and extraordinary world of meteorites. Here are just a few highlights:

- Meteorites brought water to planet Earth. Life would be impossible without them.
- Meteorites are the oldest rocks in the Solar System. Almost 4.6 billion years old in fact!
- We have meteorites to thank for all the gold available on Earth.
- Meteorites contain tiny pieces of the stars that lived and died before our Sun was born.
- You will have meteorites in your house, on your roof, and in your garden. They can be collected, it's not easy, but it can be done... This book will reveal all!

Richard Greenwood is a Senior Research Fellow at the Open University, UK. A geologist by training, he started studying meteorites quite by accident after answering an advert in a science magazine. Life has never been quite the same since. Some of his close encounters with extraterrestrial rocks are described in this book. In particular, he was the first scientist to identify the Winchcombe meteorite. But things haven't always gone totally to plan. Failure to find a single space rock on a Moroccan expedition in the 1990s still hurts. Full disclosure is provided here. Richard's more academic activities involve the study of meteorites with the aim of understanding the origin and early evolution of our Solar System. Spacecraft are now collecting material directly from asteroids and bringing the samples back to Earth. Richard has been at the heart of these activities, working with international teams analyzing material brought back by the Japanese Space Agency (JAXA) from asteroid Ryugu and by the NASA OSIRIS-REx mission from Asteroid Bennu. He also maintains a popular outreach blog on meteorites: https://meteoritestheblog.com/. When not studying space rocks, he enjoys family walking holidays in Cumbria, Scotland, Ireland, and France, regularly takes part in his local parkrun, and tries his best to keep his garden in some sort of shape. His wife thinks he needs to concentrate a bit more on that last activity.

A Meteorite Killed My Cow

Stuff That Happens When Space Rocks Hit Earth

Richard Greenwood

CRC Press
Taylor & Francis Group
Boca Raton London New York

CRC Press is an imprint of the
Taylor & Francis Group, an **informa** business

Designed cover image: Shutterstock_1042136158 and Shutterstock_1613626498

First edition published 2024
by CRC Press
2385 NW Executive Center Drive, Suite 320, Boca Raton FL 33431

and by CRC Press
4 Park Square, Milton Park, Abingdon, Oxon, OX14 4RN

CRC Press is an imprint of Taylor & Francis Group, LLC

Library of Congress Cataloging-in-Publication Data
Names: Greenwood, Richard (Research fellow), author.
Title: A meteorite killed my cow : stuff that happens when space rocks hit Earth / Richard Greenwood.
Description: First edition. | Boca Raton, FL : CRC Press, 2024. |
Includes bibliographical references and index. |
Identifiers: LCCN 2023033295 | ISBN 9781032006055 (hardback) |
ISBN 9780367774486 (paperback) | ISBN 9781003174868 (ebook)
Subjects: LCSH: Meteorites—Popular works. Classification: LCC QB755 .G738 2024 |
DDC 523.5/1—dc23/eng/20231208
LC record available at https://lccn.loc.gov/2023033295

ISBN: 9781032006055 (hbk)
ISBN: 9780367774486 (pbk)
ISBN: 9781003174868 (ebk)

DOI: 10.1201/9781003174868

Typeset in Times
by codeMantra

Contents

About this Book

Meteorites are often considered to be exciting and exotic space junk that you only ever come across in museums. The reality is very different. Meteorites are generally harmless, with the notable exception of a dead cow in Venezuela and a large number of extinct dinosaurs. Meteorites arrive on Earth every day, everywhere, usually in the form of fine dust. But bigger chunks hit the planet too. A lot of them just drop into the sea, and that is more or less it. But on a regular basis, meteorites fall on dry land, sometimes close to or even on top of people. It can be a bruising encounter, as a lady in Alabama unfortunately discovered back in 1954.

This book is really about our relationship with space rocks. You might not think you have any link with such out-of-this-world material, but you absolutely do. So, here is just one interesting fact to get things rolling: without meteorites, there would be no gold to be found on Earth. All the gold in our rings, bracelets, necklaces, etc., was delivered to the Earth by meteorites. In fact, and much more crucially, we owe everything to meteorites because without them, we simply wouldn't be here! If the dinosaurs hadn't been wiped out by a meteorite, it is very likely that mammals like us wouldn't have got a look in.

In this book, you will be introduced to the ever amazing and sometimes bewildering world of space rocks. It will show you that extraterrestrial samples arriving on Earth are not just intriguing, but they are some of the most important objects you will ever encounter. And here's the thing: people all over the world come into contact with them every day. Most of us don't even realize it, of course, but you will after you have read this book. So, first off in chapter one, we take a look at that big question: did a cow in Venezuela really get clobbered by a meteorite? You may think it's a silly question, but you will be surprised by the answer. I promise!

Acknowledgements

I have received help and assistance from colleagues, family, and friends throughout the course of this long project, for which I am very grateful. I would particularly like to thank my family: my wife, Marie-Christine, and children, Hélène, Estelle, and William, who have had to endure so much during the time I was writing this book. Stalking cows in Derbyshire and ducks in the Lake District were just two of many memorable escapades. They put their foot down at a 250-mile roundtrip to photograph the Wold Cottage monument. I don't blame them!

I am indebted to Tom Burbine who read through early drafts of various chapters, and as if that wasn't enough, then read both the unformatted and typeset versions of the entire manuscript. His encouragement and incisive comments were incredibly helpful.

I would particularly like to thank Kevin Righter, Andrew Morse, Andy Tindle, Ashley King, Beda Hofmann, Brigitte Zanda, Cari Corrigan, Carl Agee, Cathryn and Rob Wilcock, David Kring, Fiona Pepper, Graham Ensor, Ian Franchi, John Bridges, Katie Joy, Kim Gillick, Luigi Folco, Mark Errington, Martin Suttle, Mike Zolensky, Natasha Almeida, Paul Buchanan, Philipp Heck, Romain Tartese, Ross Findlay, Sara Russell, Steven Goderis, Susanne Schwenzer, Tim McCoy, Vincent Deguin, and Vinciane Debaille, for providing feedback and detailed comments on draft chapters, and in many cases, for giving me permission to use their photographs and images. This book was immeasurably improved by all the helpful advice they provided.

I am especially grateful to Naoya Imae and Akira Yamaguchi for generously giving me the benefit of their time, expertise, and insight when I visited the incredible National Institute of Polar Research meteorite collection in Tokyo. I would also like to thank them for allowing me to photograph some of the many historic specimens in the collection. Motoo Ito is thanked for his insights and help with all aspects of asteroid sample return.

Finally, I would like to express my appreciation and thanks to Rebecca Hodges-Davies and Danny Kielty at Taylor & Francis for their support, assistance, and encouragement throughout the course of this project. I am particularly grateful to Danny for his regular and gentle email nudges which kept things on course and moving in the right direction. Rebecca is thanked for her patient oversight and for her understanding when progress fell far short of my generally overoptimistic promises. I would also like to thank Karthik Orukaimani and his team at Codemantra for their efficient and professional help with the final editing and production of this book.

Richard Greenwood

1 Dead Cow

Was a cow really killed by a meteorite in Venezuela in 1972? And if so, why did the owner eat the evidence? And why did it take 30 years for the news to reach the outside world? In this chapter, we look at the facts surrounding this bizarre event and try to evaluate what it might mean for the future of humanity, if anything!

There is always so much to worry about in life: prices, illness, car problems, unemployment, visits to the dentist, getting older, not getting enough exercise, a job promotion or lack of it, a pay rise or lack of one, the weather, where to go on holiday, and so on and so forth. It seems we spend our entire adult lives worrying about something or other. So, is this book going to add another item to that list, the possibility of being hit by a meteorite? Sounds a bit farfetched and it probably is. As far as we know, the only thing that has ever been killed by a meteorite since the dinosaurs was a Venezuelan cow back in 1972. And let's be honest, even that incident is problematic, as we shall see. At this stage, perhaps we should start out by saying that death by meteorite is unlikely and there is no need to add it to your worry list or change your insurance policy, at least for the moment.

First of all, let's think about cows (Figure 1.1). Well, I don't know about you, but I have never spent much time thinking about cows! They are just sort of there and that's it. Of course, they are important, there's no doubt about that. The western diet is heavily dependent on them to provide the raw material for all sorts of food items, including the very basic one, milk. And there are lots more besides: butter, cheese, yogurt, custard, cream, crème fraiche, ice cream, cottage cheese, buttermilk, etc. (Figure 1.2). That's a lot!

Now to be honest, I am being a bit disingenuous saying that I have never spent much time thinking about cows. Once upon a time, I was very keen on them. My Mum was born and brought up in a very rural part of County Tipperary, Ireland. Actually, almost all of County Tipperary is rural. There are a few towns, but most of it is fields, very green fields. If you get a chance, it really is worth a visit.

Every summer when I was growing up, we would head off to stay on the family farm for a two-week holiday. It was a small dairy farm of about 100 acres, split between two sites, each with a small herd of Friesian cows, the black and white ones,

DOI: 10.1201/9781003174868-1

FIGURE 1.1 Not the actual cow killed by the Valera meteorite, not even the same breed, but you get the idea. Is it really possible that a Venezuelan cow was the only victim of a falling meteorite? The cow in the photo is from Derbyshire, UK and doesn't seem particularly troubled by the very remote possibility of being hit by a meteorite. (Photo: the author.)

which a giant ice cream company now uses to promote its products. They were milked by hand in the morning and evening by Uncle Matt on one site and Uncle Desmond on the other. Milking machines may have been catching on elsewhere in the world, but in this gentle backwater of rural Ireland, things were still done the traditional way. As children, we were fascinated by the milking process. The cows were led into the stalls in the dark, stone barn, with my uncles invariably sitting on old, very low, three-legged stools. They would use their enormous hands to rhythmically milk each cow in turn. To us, the cows were giants, and of course, they are pretty big animals, but more of that in a bit. They are also quite docile, except now and then when one of them would start getting a bit frisky. This was quickly sorted out by a gentle slap on the back, and at the same time, a deep, loud rebuke would be given, along the lines of HEYOO, or something like that. We had no idea what it meant, but the cows did. It was certainly a different world to the modern-day, high-tech milking parlours that provide us with a seemingly inexhaustible supply of milk. Once the milking bucket had been filled with warm frothy milk, it would be emptied into a big metal churn to join the previous evening's takings. Later in the morning, the churn was loaded onto a trailer and driven to the creamery in Thurles to be turned into a range of products,

FIGURE 1.2 There's no doubt about it, we owe a lot to the humble cow! (Photo: the author.)

pasteurised milk of course, but also butter, cheese, ice cream… (OK that's enough, we went into all that earlier.) And from time to time, we got to travel on the milk run. It was the highlight of the holiday. Not because we were that interested in seeing how milk gets converted into a range of varied, derivative products, but because we got to sample them. In particular, the ice cream. The Thurles creamery made an amazing chocolate-coated vanilla ice cream that tasted like heaven. Well, we thought so at the time. The creamery is still going strong today, but alas they no longer make those amazing ice creams. Perhaps we can get a campaign going to bring them back; they were delicious.

In summary, to say that I have never thought much about cows is not strictly speaking true, I just haven't thought about them much recently. To realise that they are the only known victims of death by meteorite deserves a little more scrutiny. So, here we go.

The dead cow in question is said to have been killed by a space rock that is now officially known as the Valera meteorite (Figure 1.3).

FIGURE 1.3 A 144g slice of the Valera meteorite. Valera is an ordinary chondrite type L5. That just means it is one of the most common types of meteorites arriving on Earth. But don't worry too much about the technical stuff for the moment, we will get to all that shortly. (Photo: Graham Ensor.)

To find out more, we need to take a look at the official entry for Valera on the amazing Meteoritical Bulletin (Met Bull) Database [1]. It is the world's single most authoritative resource on individual meteorite samples and is hours of fun! The entry for Valera tells us that it is a confirmed "fall" [2] that took place on 15th October 1972 in Trujillo, Venezuela. It says that the mass of the meteorite was 50kg and that it was an "ordinary chondrite", type "L5" [3]. We don't need to worry about that technical stuff for the moment (see Appendix 2 for further information). Right now, all you need to know is that the Valera meteorite is a very hard and heavy lump of space rock. The Met Bull Database provides some more details about the events that surrounded its fall:

> On the evening of 1972 October 15, a bright light accompanied by a loud noise was witnessed near the El Tinajero farm. The next morning, Dr. Arginiro Gonzales and his guest, Juan Dionicio Delgado, discovered that a cow had apparently been killed by a falling stone. The stone had broken into three pieces weighing 38, 8, and 4kg, respectively. The largest specimen remained outdoors for decades after the fall. Classification and mineralogy (A. Rubin, UCLA).

The last part of the entry tells you that Dr Alan Rubin of the University of California, one of the world's leading meteorite experts, studied the sample and provided the data required by the Nomenclature Committee (NomCom) [1] to make an official verdict on whether Valera represented a genuine space rock. There can be no doubt, on the basis of the information provided by the Met Bull Database, that Valera is a genuine meteorite. But did it kill a cow? Of course, the NomCom is not going to give an opinion on that. It just reports the information that it has been given about the circumstances of the fall. More details on Valera are provided by a 2006 Meteorite Times article by Martin Horejsi [4]. It seems that although Valera fell in 1972, not much happened until the very early 2000s, when it was officially classified. At that time, an affidavit dated 11 January 2001 was signed by Juan Dionicio Delgado in which he relates the events that had taken place back in 1972. The owner of the farm Dr Arginiro Gonzales had unfortunately died sometime before the signing of the affidavit. The official English translation of the affidavit makes interesting reading:

> I, Juan Dionicio Delgado, Venezuelan, identified by the National Identity Document No. 5.030.450, by the present document I declare that at the end of 1972 I was visiting the farm "El Tinajero" owned by Argimiro Gonzalez, deceased, which was located in the limits of the states Barinas and Tujillo. It was past midnight when we were talking, and there was a strange noise. When we went out to investigate due to the dark of night we saw nothing. But the next morning a worker came to say that there was a cow killed in estrange circumstances. When we went to investigate we found the cow had been killed by a stone that presumably had fallen from the sky the night before, causing the noise that we had not been able to explain. The stone that had been broken in several pieces, was kept by Dr Gonzalez, while the cow was eaten in the following days. History that I relate in Barinas, eleven days of January, 2001.

From the affidavit, we learn that the unfortunate cow, having been killed by a falling space rock, was shortly thereafter eaten. I suppose that the cow wasn't too bothered and it made some sort of sense to eat it and not let the carcass go to waste. On the other hand, we get no details in the affidavit about what sort of state the animal was in after being killed by a flying 50kg meteorite. It also seems a bit surprising that it took so long for this amazing story to come to prominence.

Interest in the Valera meteorite has grown since it was first officially classified back in 2001. This is not surprising in view of the reported circumstances of the fall. Like all rare natural objects, meteorites not only have an intrinsic scientific value, but also have a financial value. As we will learn later in this book, there is an active trade in extraterrestrial rocks and Valera is no exception. A 160g piece of Valera came up for auction in April 2016 at Christie's of London and achieved a sale price of £5,250 [5]. Meteorite prices are often quoted in dollars per gram, so if we accept a dollar value for this sale of $6,000 (exchange rates do fluctuate a bit), that makes a price per gram of $37.50, which is actually a relatively modest amount for a meteorite with the back story of Valera. The Christie's sale documentation includes a "lot essay" which adds new details to the Valera legend:

> An echo in miniature of the devastating asteroid believed to have killed off the dinosaurs, it was on the evening of October 15, 1972 that farmhands in Trujillo, Venezuela were startled by an inexplicable sonic boom. The next day an exotic rock was found alongside a cow's carcass whose neck and clavicle had been pulverized. It was obvious to the farm's owner, physician Dr. Argimiro Gonzalez, what had occurred, but he didn't give it a second thought since mayhem from falling meteorites seemed intuitive. An unplanned steak dinner was enjoyed that night and the celestial boulder was used as a doorstop. More than a decade later scientists confirmed what Dr. Gonzalez had long presumed. However, what Dr. Gonzalez didn't know was that this was the first and still the only documented fatal meteorite impact. When Dr. Ignacio Ferrin, an astronomer at the University of the Andes, learned of the act of bovicide that had occurred at Valera, he visited the Gonzalez estate and left with an affidavit affirming the aforementioned events as well as the meteorite itself.

Also in the Christie's sale particulars that accompanied the auction, it was made clear that the affidavit from Juan Dionicio Delgado was an important supporting document stating: "*Accompanied by a copy of the signed affidavit attesting to the circumstances of the Valera event*". And where did the meteorite sample that was on sale come from? The auction particulars state that it was from "Dr. Ignacio Ferrin, Merida, Venezuela". Dr Ferrin was perhaps the source of the details about how the unfortunate cow died, namely a pulverised neck and clavicle. In other words, the cow's neck had been broken, a fairly restrained sort of injury, which might explain why everyone seemed happy to eat the carcass after it was discovered. A totally shredded cow would presumably have been a less appetising prospect.

A slightly smaller 95.1 g piece of Valera came up for auction with Christie's in the US in February 2021 and fetched a much higher price than the earlier UK sale [6]. The material went for $20,000 making it a relatively respectable $210 per gram. A lot of factors could have contributed to the increasing financial value of the Valera meteorite, but one possibility is that amongst collectors, it had started to gain a certain notoriety as the cow-killing meteorite.

But should we take this cow-killing story at face value? In an article written for the Guardian newspaper in January 2002, Duncan Steel expressed a note of caution [7]. He pointed out that stories about animals being killed by meteorites were nothing new. The most famous example was the reported death of a dog in Egypt in 1911, which it has been claimed was dispatched by the scientifically important Nakhla meteorite. In a 2011 article for the Smithsonian website, marking the 100th anniversary of the Nakhla meteorite's fall, Arcynta Ali Childs points out that the Nakhla dog-killing claim cannot be substantiated [8]. However, she does describe the details given by a local farmer as "irresistible". Here is what he said:

> The fearful column which appeared in the sky at Denshal was substantial. The terrific noise it emitted was an explosion which made it erupt several fragments of volcanic materials. These curious fragments, falling to earth, buried themselves into the sand to the depth of about one metre. One of them fell on a dog…leaving it like ashes in a moment.

By the way, Nakhla is a Martian meteorite, a chip off the red planet [9]. But more of that later. There is also the disputed case of the horse killed by the New Concord meteorite in Ohio in 1860, for which there appears to be no contemporary evidence [10]. Duncan Steel is clearly a little sceptical about the cow killer claims for Valera and rather mischievously suggests that there were commercial motives involved in the reported events surrounding its fall. He speculated that the bovine death claim may have been motivated by the possibility that it would help to enhance the sale value of the meteorite. And if this was the case, he had a suggestion about where the inspiration for the conspiracy might have come from.

Vaca Muerta is a huge meteorite found in the Atacama Desert of Chile [11]. It was discovered in 1861 and has a total recovered mass of 3.83 tons. Interesting! But what has that got to do with Valera? Well, nothing in terms of its composition, the two meteorites are completely different [12]. But Duncan Steel says the connection may be in the name, as Vaca Muerta in Spanish means "Dead Cow". He points out that some "cynics" have suggested that this name may have given the people behind Valera the idea of linking it to a cow-killing story in order to increase its financial worth. A very Machiavellian scenario! Is it a smoking gun? Have we uncovered a conspiracy to defraud the extraterrestrial community? I don't think so. It is really just an interesting speculation and, without additional evidence, doesn't help us to decide whether the Valera cow-killing story is true or false.

Another approach is to take the Valera recovery story at face value and see whether it makes sense based on what we know about the behaviour of meteorites as they pass through Earth's atmosphere. Valera had a total recovered mass of 50 kg. It is a tough piece of rock and, based on the normal characteristics of similar meteorites (Figure 1.4) [13], would have had an approximate diameter of about 32 cm when it hit the cow. In other words, it was essentially just a large boulder. When they arrive at the top of the Earth's atmosphere, meteorites are travelling at speeds of between 25,000 and 160,000 mph (11–72 km/s) [14]. But they rapidly slow down as they penetrate deeper and deeper into the atmosphere [14]. At the end of their flight, meteorites the size of Valera reach "terminal velocity" and will be travelling at between 200 and 400 mph [14]. Perhaps that's not as fast as the crazy speed it would have been doing before entry, but it's still not bad. Just imagine the potential damage a 50 kg boulder travelling at 200 mph could do if it hit a cow!

Is it credible, based on what are admittedly very sketchy details, that the Valera cow was relatively intact after this encounter, apart from a "pulverised" neck and shoulder of course, and the fact that it was dead? Surely, its injuries would have been much more severe. And would anyone really have wanted to eat the leftovers after such a high-speed encounter? On the basis of the patchy details provided, the cow-killing story starts to seem a bit farfetched, to say the least. Perhaps, as hinted at in the Guardian article, it was all made up to improve the sale price.

But then I had another thought…

Maybe the Valera cow wasn't hit by the meteorite during its initial impact at all, but rather during some sort of secondary event. Did the Valera meteorite behave like an old-fashioned cannonball and bounce?

In a celebrated accident that took place on the American TV show Myth Busters, a cannonball fired from a mock-up of an old-fashioned cannon went astray, bouncing

FIGURE 1.4 This is not the Valera meteorite, but like Valera, it is an ordinary chondrite (see Appendix 2 for further details). As would have been the case for Valera, it has a thin, dark outer layer called fusion crust. In a few places, the fusion crust has broken off and you can see that the interior of the stone is much lighter in colour. The stone also shows well-developed "regmaglypts" which are shallow, thumbprint-sized depressions formed by non-uniform "erosion" of the meteorite's outer surface, as it was heated during its flight through the atmosphere. The meteorite shown here is known as Northwest Africa (NWA) 869 and was part of a large meteorite that broke up in the atmosphere before impact. As far as we know, it didn't kill any cows…. (Photo: the author.)

numerous times and passing through a series of properties and cars before finally coming to rest [15]. Happily, no one was hurt in the incident, but this illustrates the fact that a compact solid object flying at high speed, such as a cannonball (or a meteorite?), doesn't just come to rest instantaneously. And this wasn't unusual behaviour for cannonballs. They were designed to bounce and cause mayhem on the battlefield [16]. Cannonballs of old, fired from large, classic, heavy cannons, the sort of thing you see displayed at any self-respecting ancient monument (Figure 1.5), are more technically a type of "round shot". We may think of them as being quaint and cuddly objects, useful as paper weights and suchlike. But think again. When fired in anger, they were devastating [16]. They could fly through massed ranks of troops causing multiple casualties. And they didn't stop nicely and abruptly on first striking the ground. They were high-velocity bowling balls and would continue on bouncing and hitting men as they went. A single shot could kill and maim large numbers of troops and the injuries were horrendous, even for those that encountered the projectile during the later stages of its trajectory [16].

Fine, but could a meteorite really behave like a ricocheting cannonball? It turns out that this is not an uncommon event. In a review of oblique meteorite impacts, Pierazzo and Melosh [17] point out that "*natural impacts in which the projectile*

FIGURE 1.5 Cannons outside the Tower of London. These may look cute and old-fashioned now, but they were designed to cause mayhem on the battlefield. The cannonballs (round shot) that they discharged were designed to bounce and cause very severe injuries. (Photo: Estelle Greenwood.)

strikes the target vertically are virtually non-existent". In other words, most incoming space rocks do not just drop vertically from the sky at high speed. These authors go on to discuss the results of experimental simulations, which show that, as the angle between the flight direction of the incoming space rock and the ground reaches about 30°, projectile ricochet becomes the norm. It is very much what happens when you skim a stone across the surface of a pond. Pierazzo and Melosh [17] also have this to say about the fate of the incoming space rock or "projectile": "*depending on the projectile strength, and with increasing impact velocity, ricochet may occur either with the projectile remaining intact, rupturing into several large fragments, or shattering into a myriad of small fragments*". It's not proof of course, but Valera, if you remember, was found broken into three fragments weighing 38, 8, and 4 kg. Perhaps this wasn't due to an impact with the cow, but a consequence of a nearby glancing blow with the Earth. Perhaps just one of these fragments killed the cow. But of course, we will never know. Experimental studies, which usually involve firing solid objects into buckets of sand, have shown that projectile ricochet is a common event on small and large asteroids, and even on planets [17,18]. Intriguingly, these studies suggest that if an incoming space rock is travelling at a high-velocity and at a low-angle trajectory, it could hit the Earth and then bounce back into space [17]. I wonder how many times that has happened? Worth a thought next time you skim stones across a pond.

And finally, perhaps we should look at this from the point of view of the cow. There are about 8 billion people on this planet at present and, according to the United Nations Food and Agriculture Organisation [19], about 1.5 billion head of cattle. That's close to five people for each head of cattle and yet it is the cows, not the people, who have been singled out for extraterrestrial death. It doesn't seem fair. Hold on! We are only talking about one cow here and they are big animals. The body mass of an average human varies from 57.5 to 80.7 kg [20], whereas the Criollos – Spanish Longhorn cattle that are commonly reared in Venezuela come in at an average weight of 362 kg [21]. That's 4.5–6.3 times heavier than an average human. Since humans outnumber cows 5:1, you would think that as many humans as cows should have been taken out by space rocks. But there are a number of other factors to take into consideration. As we have seen, cows are bigger targets than humans, they also spend longer outdoors, and in general, are more sedentary. There are local factors at play too. Venezuela has a lot of cattle. The current human population of Venezuela is 28 million and the cattle population, as estimated by the US Department of Agriculture [22], is 15.4 million. That makes for more than half a cow for every person. Of course, things could have been different back in 1972. But if anything, reports suggest that Venezuelan cattle numbers were higher back in those days. It seems that from a statistical point of view, the killing of a cow by a meteorite was not a totally unreasonable outcome.

But it was just one cow, and let's get real here: there is no compelling evidence that the event ever took place. It's frustrating. So, finally, a heartfelt plea to landowners globally. If you suspect that one of your animals has been wiped out by a meteorite, please don't eat the evidence and then wait nearly 30 years to tell the world about it. Do contact your local planetary scientist and let them know about the incident. We would really like to hear about these sorts of space rock-bovine encounters, sooner rather than later!

Now it's time to move on and learn about a human meteorite interaction in which an elderly gentleman was almost wiped out by a flying space rock. We are off to England and the lovely, well-manicured garden of Mr Arthur Pettifor.

2 Mr Pettifor's Garden
A Very English Meteorite Adventure

Here we find out about a meteorite that nearly killed a man in his garden and then went for a ride on the London Underground. You just couldn't make it up!

It was a grey, overcast day in southern England, just before lunchtime on Sunday 5 May 1991 [1,2]. Mr Arthur Pettifor was busy working in his garden in the village of Glatton, Cambridgeshire. Mr Pettifor always kept his garden in perfect condition. There were neat flower beds and straight rows of vegetables. Nothing, absolutely nothing, was out of place. In short, Mr Pettifor's garden was a magnificent example of British domestic horticulture. As he worked in his garden that day, sorting out his onions, Mr Pettifor was startled to hear a loud whining noise. It was almost immediately followed by the sound of splintering wood, as something hard, travelling at a fair old speed, hit the branches of the conifer trees at the bottom of the garden [1,2]. As we know from the Valera encounter, meteorites are still travelling at several hundred miles per hour when they hit the ground, even despite being slowed significantly by atmospheric drag forces. Although he may not have realised it, Mr Pettifor was a lucky man. Had he been hit by the space rock, it would almost certainly have killed him.

A report on the Glatton Parish Council website (now deleted, alas) suggested that the noise was also heard by a nearby neighbour in High Haden Road and that the two men then went to look at the damaged conifers together. Mr Pettifor had noticed from afar that one of the branches of his tree was slightly damaged by the impact. When he reached the bottom of the vegetable patch, he was surprised to find a small angular, black rock lying in a shallow depression at the base of the trees. He picked up the rock. It was slightly warm to the touch, had a dark, shiny exterior, and measured about 10 cm × 6 cm × 6 cm. Mr Pettifor knew instantly that this rock was out of place in his orderly garden (Figure 2.1).

Shortly afterwards, and in true British tradition, Mr Pettifor, would no doubt, have invited his neighbour in for a cup of tea. This is speculation on my part, but after such an event, a nice cup of tea would surely have been de rigueur in this tranquil corner of England. But what would the two men have made of the mysterious stone?

DOI: 10.1201/9781003174868-2

FIGURE 2.1 Mr Arthur Pettifor holding the Glatton meteorite. To see the Glatton meteorite for yourself, check it out on the Virtual Microscope: https://www.virtualmicroscope.org/content/glatton. (Photo: © The Trustees of the Natural History Museum, London.)

Later, local reports claimed that from the start, both were convinced that the flying rock was a meteorite. But perhaps, for a second or two, they might have considered the possibility that someone had lobbed the rock over the fence? But no! Such things didn't happen in Glatton.

Then again, there was always the chance it could have been something that had dropped off an aeroplane. Unfortunately, pieces of debris descending unexpectedly to Earth from planes are not as rare an event as you might think. There are even two acronyms for the problem: "PDA" – Parts Departing from Aircraft and "TFOA" – Things Falling Off Aircraft [3]. Happily, in most cases, these incidents do not result in serious consequences. In July 2014, a 60 cm diameter external diffuser fan fell from an aircraft departing Chicago's O'Hare International Airport and landed in an empty children's splash pool at the Bensenville Waterpark [4]. In other incidents, nuts and bolts and even pliers left by maintenance engineers have been known to drop to Earth from planes. And then there are encounters with the so-called "Blue Ice", which consists of a mixture of human waste and liquid disinfectant that has leaked from an aircraft's sewage system and become frozen at high altitude [5]. You wouldn't think that people would mistake such material for meteorites, but they do! In 2018, the BBC reported the fall of a 10–12 kg block of ice in a village in the northern Indian state of Haryana [5]. A local official, Vivek Kalia, was reported as saying:

> "It was a very heavy icy ball of ice which dropped from the skies early on Saturday morning. There was a big thud and people of the village came running out of their homes to find out what had happened. Some villagers thought it was an extraterrestrial object. Others thought it was some celestial rock and I've heard that they took samples home".

It turned out to be "Blue Ice"!

But Mr Pettifor and his neighbour hadn't found a piece of aircraft-related debris, and in due course, their conviction that they had been visited by a meteorite would receive an official endorsement. Not from me I should stress, but I did play a small role in the enfolding drama. It's time to pay a visit to one of the UK's great scientific institutions, The Natural History Museum in London (Figure 2.2).

A year before the arrival of Glatton, after answering an advert in the New Scientist magazine, I landed a job in the Mineralogy Department of the Natural History Museum, working on, you guessed it, space rocks. To be honest, at the time I got the job, I knew close to nothing about meteorites, but figured that they were just rocks, and while they had travelled a bit further than the ones I had been studying in Scotland, well, a rock was still a rock! Anyway, I thought it was worth a shot. And in due course, I was majorly surprised and very pleased, when I got an interview for the post. I headed up to London to meet the "Curator of Meteorites", Dr Robert Hutchison. It was a small world because I had actually met him a year or so earlier in a castle on the Scottish island of Rum, where I had been doing some fieldwork for my PhD. Robert Hutchison was a very friendly man, but also a Glaswegian, who said what he thought and didn't beat about the bush. I was a bit worried about the interview because I had this vague recollection that our earlier encounter hadn't gone so well. Perhaps a little bit too much Scottish mist had been imbibed at Kinloch Castle that particular evening. During the job interview, we spent a lot of time talking about Scottish geology and I was relieved to find that there seemed to have been no lasting damage from our previous meeting. For some reason that escapes me now, as the interview continued, I started to relax a little too much. With reference to the post I was applying for, I let slip the following daft question: "*Does it matter that I don't really know that much about meteorites?*" I am not sure what I was expecting him to reply. "*No problem, we are more than happy to appoint someone who admits that they have absolutely no subject knowledge relevant to the position they are applying for*". And of course, he didn't actually say that. He looked at me with a slightly puzzled expression and said "*Well, a little bit of knowledge would probably be helpful*", or something along those lines. So, there you go! That was one job I wasn't going to get. Nice try! In fact, not a nice try. How to blow things out of the water with a single question. At least that's

FIGURE 2.2 The Natural History Museum, London which has one of the world's finest meteorite collections and where I worked at the time the Glatton meteorite landed in Mr Pettifor's garden. (Photo: Estelle Greenwood.)

what I thought as I headed back to Hampshire on the train. But for some strange reason, Robert Hutchison was able to see something positive in our second encounter and surprisingly decided that I was the man for the job. I don't know what the reason for his slightly baffling decision was, but I remain ever grateful.

Just about one year on from that interview, I was sitting in my office next to the Cromwell Road, basically in a basement storage area where all the Museum's meteorites were kept, and still are, even to this day. It was an amazing place to work. One of the world's finest collections of space rocks was kept in large wooden cabinets, literally two steps from my desk. Robert Hutchison's office was directly next door to mine and on that particular day, his phone just rang and rang. He was clearly not around, so finally I answered it. As one does. "*Hello, is that Dr Hutchison?*" "*No, I'm afraid he's out, can I take a message? I am his researcher.*" Not sure that's exactly what I said, but something like that anyway. Well, it was a few years ago. The person at the other end of the line had been contacted via a string of people, initiated it seems by a member of the local astronomy club. Based on the published account written by Robert, the man on the phone must have been Howard Miles of the British Astronomical Association. He said something along the lines of "*Could you tell Dr Hutchison that a bloke in a village near Peterborough had a rock land in his garden and we think it is the real thing*" I took the details, telephone number, etc., and thanked him for getting in contact. Later that day I gave Robert the message. Of course, it sounded interesting, but we fairly regularly got calls about mysterious space rocks arriving in people's gardens. The most notorious was before my time at the Museum. Apparently, someone had given Robert a textbook description of a meteorite over the phone, but when he turned up at their house to study the specimen, it was a brick. Not a disguised, battered brick that could be mistaken for a meteorite, if such a thing exists. No, it was a brick that looked like a brick! (Figure 2.3).

Then there were almost daily enquiries about small rounded "concretions" found in the local chalk rocks. They do have a sort of meteorite look about them, it has to be said. We were always happy to examine what the general public brought in, after all, that was part of our job. But with so many false alarms, I wasn't holding out a lot of hope for the stone from Glatton. But I was wrong. When Robert came into my office the next day, he was very upbeat. He had been up the evening before to visit Mr Pettifor and examine his rock. I think Robert knew from the very start that this was the real deal and now there was no longer any doubt about it. The UK had a new meteorite fall.

Now things started to move pretty swiftly. A meteorite fall is a rare event and the media were up for it. Mr Pettifor, the man who found a meteorite at the bottom of his garden, became the true star of the show. His photo appeared in umpteen national newspapers. He was interviewed by all the evening news programmes. But calling it a media frenzy would be going too far. After all, stories about space rocks were not as newsworthy as the day-to-day schedule of the Princess of Wales. Nonetheless, Mr Pettifor became, for a short time at least, a bit of a media star.

Several days later, as part of a diverse group of enthusiasts and students, I headed up to Glatton. Robert's account gives more precise details of this event [1]: "*On Sunday 12 May volunteers searched E-W swathes of farmland both north and south of the place of fall, looking for more stones; none was found.*" My recollection of that day was that events were a little less formal than this account suggests. We all met at Mr Pettifor's house and had a good look around his beautiful garden, where all the action had taken place a week earlier. We saw the broken branches and the depression in the soil the meteorite had made as it came to a final halt. It was all very exciting.

I can't remember now when I first saw the Glatton meteorite. It may have been at the Museum on Friday 10 May, the day after it had been identified by Robert. I am pretty sure he brought it to the Museum that day with Mr Pettifor's permission. Alternatively, I first viewed the space rock at Mr Pettifor's house on that Sunday morning, along with the other meteorite aficionados. I do remember being astonished

FIGURE 2.3 One of the most celebrated "meteorwrongs" dealt with by the Museum was a brick that looked like a brick. Identifying genuine meteorites is not easy. Appendix 1 gives some tips on what to look for. But if it looks like a brick, feels like a brick, and has "London Brick" engraved on it, then it might not be a meteorite. (Photo: the author.)

when I first saw it. I had never seen anything quite like it before. Yes, we had loads of meteorites in the Museum's collection, but this one was different. It was fairly small, blocky, a bit angular, but with soft rounded edges [6]. A corner of the specimen had broken off due to a domestic mishap – someone had dropped it on the kitchen floor, I believe. But that was an advantage in a way. Now you could see that this dark black shinny exterior was just a paper-thin layer, and on the inside, the meteorite was much lighter in colour, almost white in fact. I knew from looking at other fresh meteorite samples that this black outer coating is always present on newly fallen meteorites. As we saw for Valera, it is known as fusion crust [6] and forms due to melting of the outermost part of the stone as it is heated by friction, while descending rapidly through the Earth's atmosphere.

Next, we fanned out into the local countryside and looked for more bits from the possible meteorite "shower" [7]. We trudged around for a good few hours. The countryside around Glatton is very rural and pleasant. But gradually, we started getting a little bit frustrated and fed up. We found nothing. Inevitably, a drift in the direction of the local pub started to take place. Apparently, it served an excellent pint. I did carry on with the search for a little longer, but finally gave up as well. To this day, no further pieces of the Glatton meteorite have ever been found. They may be out there just waiting to be picked up. Meteorites don't usually stay in one piece as they plunge through the Earth's atmosphere [7]. The forces grow on the descending stone as the atmosphere gets thicker and denser. Finally, they break up and land on Earth as a shower of stones. It seemed to us at the time that there was a reasonable chance other bits of Glatton might be lying in the fields around the village. Perhaps these bits were bigger than Mr Pettifor's stone. It was not uncommon (as we shall see) for an initial isolated piece of space rock to be located, and then following a hunt of the local area, many more stones are recovered. However, although we didn't know it at the time, the results of the next development in the Glatton story made the chances of more chunks of it being found unlikely, but not impossible. If you are in the Glatton area, it is still worth keeping an eye out!

The next day, on Monday 13 May, Robert was back in the Museum along with the meteorite. It was now on extended loan for scientific measurements to be made. In passing, Robert mentioned that he was off to catch the Tube to Holborn underground station and would be taking the meteorite with him (Figure 2.4). I thought this was a bit odd. Why would you take a precious, newly fallen extraterrestrial rock for a ride on what is a far from clean mass transit system? Was it perhaps some sort of weird, Natural History Museum tradition, thingy? Newly acquired samples are taken for a swing around the bowels of London as a kind of initiation rite. There was a lot of strange stuff going on at the Museum. A wonderful scientific institution but sometimes a bit of a bonkers place to work. "Things" happened at the Museum, and it was best not to ask too many questions.

Part of the Natural History Museum is built on top of a World War II bunker, "South Kensington Home Security Region 5 War Room" to give it its full official title [8]. To make way for an extension to the Museum, an attempt was made to demolish the bunker in 1976. But its external walls are nearly 2 m thick and composed of bomb-resistant reinforced concrete. It wouldn't budge. After all attempts to remove it failed, the decision was taken to build the new extension on top of it, turning the bunker into a useful subterranean storage area. At the time I worked there, and possibly still today, part of the bunker contained human remains from abandoned cemeteries uncovered by accident during building works throughout London [9]. It is an important scientific resource, known as the "London human remains collection" and gives archaeologists a unique insight into the lifestyles, nutrition and medical histories of the past inhabitants of London. And although I thought it was fascinating, it wasn't really my thing. I went to see it once and never went back. It was the sort of thing the Museum did.

While I was there, I also heard a lot about the Museum's "Spirit Collection", a unique store of dead animals and other organisms preserved in formaldehyde [10]. However, I gave that a conscious miss. Dogfish dissection work undertaken in biology

FIGURE 2.4 Holborn underground station where the Glatton meteorite went for a short visit. (Photo: the author.)

classes at school had been a bit traumatic, as the teacher had left the specimens pickling for almost a full academic year before we were given them. It had been an eye-watering experience!

So, based on the sort of activities that seemed to be a part of Museum's daily life, taking a meteorite for a ride on the London Underground sounded par for the course and hey! Why not? Show it a few sights. The Underground is a chill place and Holborn is a nice, well OK-ish, underground station. But Robert soon put me straight. He must have sensed my disbelief and proceeded to give me a little explanation as to why this subterranean expedition represented a valid research activity. The following is not the actual words he used, of course, but they give you the gist of what he said:

"I am taking the meteorite to John Barton's lab down in Holborn underground station. This little rock has come a long way. It started its journey buried within an asteroid in the asteroid belt, located somewhere between the orbits of Mars and Jupiter. Then one day it was kicked out into space when another space rock slammed into its home asteroid and so began a journey that ended up in Mr Pettifor's garden. Now, as a small object wandering the inner Solar System, it would have been irradiated by that giant nuclear reactor we call the Sun. As a result, it would have become mildly radioactive. But that would have changed when it plunged to Earth. Shielded by our atmosphere, this natural radioactivity would have started to decay away. By using the very sensitive equipment in the Holborn lab, we can measure this radioactivity and use it to work out how big the meteorite was before it hit our atmosphere and also how long it was

wandering the Solar System after it left its parent asteroid. And that is what I am going to do in John Barton's lab in Holborn station."

And with that he was gone. Every day at the Museum was an education.

The Holborn laboratory had originally been set up to detect neutrinos [11] (Figure 2.5). The detectors that do this are normally housed at the bottom of deep mines to shield the equipment from the interference of cosmic rays. Unfortunately, deep mines are in short supply in central London, so it was hoped that a lab 100 m below ground might do the trick. But it turned out that this was just not deep enough. However, the lab did prove to be useful for other experiments that needed to be away from London's urban noise and at least some of the influence of cosmic radiation. The labs were reached through a service door at the end of one of the Piccadilly line platforms (Figure 2.5). That day back in 1991, Robert must have seemed a strange sight as he disappeared through this narrow door holding a briefcase with Glatton inside. Perhaps a little like a scene from the British science fiction film "Quatermass and the Pit" [12], in which the London Underground plays a prominent role.

And what was the result of the work carried out in the Holborn lab? Counting was undertaken on the whole meteorite (730 g), less a 37 g piece removed for classification studies. Counting began eight days after the fall and continued for 23 days. The spectrum that was acquired indicated that in the ten years before atmospheric entry, Glatton had a diameter of less than 20 cm. In other words, it was significantly smaller than Valera. As a considerable amount of material would have been lost from

FIGURE 2.5 John Barton's lab was reached through a service door on one of the Piccadilly line platforms – a bit like this one. (Photo: the author.)

its surface during entry, it therefore seems unlikely that there could have been many other pieces of Glatton that fell on the same day as the one that landed in Mr Pettifor's garden. Glatton was small, and if Mr Pettifor hadn't been in his garden working on his onions and if he hadn't been such a meticulous gardener, it would certainly have gone unnoticed. But that begs the question, how many small meteorites like Glatton land on Earth and are never recovered? And of course, the answer must be loads.

At the same time as John Barton was making his measurements in Holborn station, Robert used the remaining 37 g piece to collect the mineral data required to get an official classification for Glatton. The results of this study showed that it was an ordinary chondrite, type "L6" [13]. That is the same group as Valera, with the "6" for Glatton indicating that it underwent slightly more heating in an asteroid than Valera did, which as you will remember is "5".

We should just pause here for a moment and add a little bit more detail, so that all this meteorite classification jargon makes some sort of sense. Scientists are lovely people, but they are very fond of using long complicated words to explain what can often be quite simple ideas. So here goes!

Glatton and Valera are "ordinary" chondrites. The "ordinary" really just means that they are the most common type of meteorite arriving on Earth at the present day. Ordinary chondrites represent about 80% of all recorded meteorite falls. There are other rarer types of chondrites, and when these are added in, chondrites represent about 87% of all falls. The term chondrite just refers to the fact that these meteorites contain "chondrules". That's right, chondrules! What are chondrules? Well, answering that question is the easy bit. What is really tough is working out how they formed.

Chondrules are silicate-rich spheres, generally up to about 1 mm in diameter, found in "chondritic" meteorites (Figures 2.6 and 2.7) [14]. Chondrules have distinctive internal textures, which experimental work suggests, formed during a rapid heating event followed by a period of relatively rapid cooling [15]. And chondrules are old. They are not the oldest Solar System objects, but we will get into all of that in Chapter 11. Chondrules formed no more than 3 million years after the start of the Solar System, which is currently dated as having come into existence 4,567 million years ago. On this timescale, 3 million years is just the wink of an eye [16].

In short, chondrite is the term used for a meteorite that contain chondrules. And some chondrites contain a lot of chondrules. Ordinary chondrites, like Valera and Glatton, can contain between 60% and 80% chondrules by volume. Working out how chondrulres formed remains a fundamental problem in meteorite science [14]. And how well are the scientists doing in sorting this out? To be fair, they have come up with a very wide range of possible processes, but as yet, there is no single theory for chondrule formation that everyone agrees on. Well, that's science for you. It's what makes it fun.

While the existence of spherical structures in meteorites had been known about from the start of the nineteenth century, the first person to put forward a theory for their origin was the Englishman Henry Sorby in 1877 [17]. His suggestion, which is still valid, was that they "*were originally detached glassy globules, like drops of fiery rain*". When chondrules were forming, the Solar System was very different from its current sedate state. The Sun was still surrounded by a rotating disc of

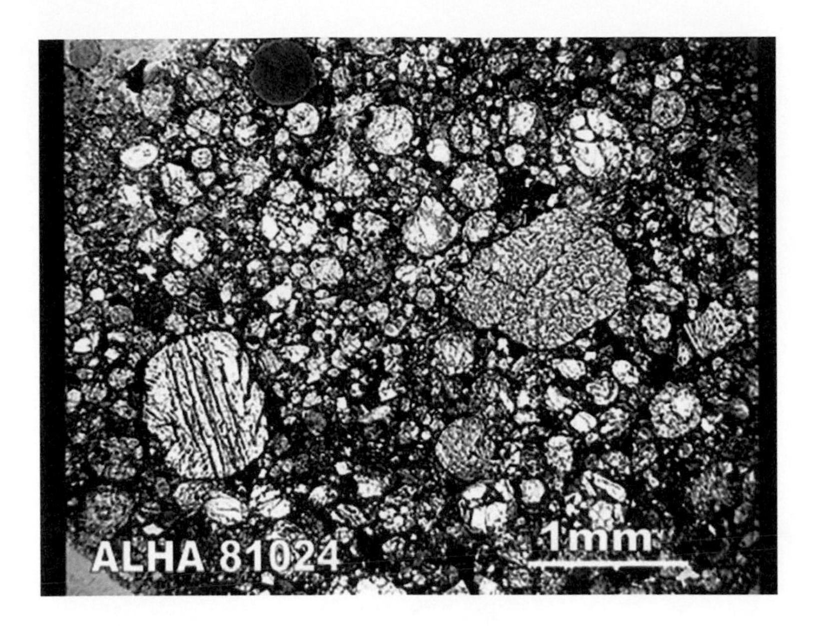

FIGURE 2.6 This is a microscope image of ordinary chondrite ALHA81024 collected in Antarctica in 1981. Like all ordinary chondrites, it is packed full of spherical objects called chondrules. The largest chondrules in this section have a maximum diameter of 1.5 mm. Chondrules are ancient and formed within 3 million years of the start of the Solar System [14,16]. Chondrules were produced by melting of dust grains in the dense cloud of material that surrounded the young growing Sun. At that time, planets like Earth and Mars had yet to form. (Image: NASA/Smithsonian.)

dust and gas. Small planetary bodies had started to form and the Sun itself was in a highly energetic state. This complex environment has led to a raft of chondrule formation models. These include: heating by lightning discharges, melting of primitive dust by shock waves, formation as sprays of molten droplets following collisional impact between partially melted mini-planets, heating of dust by energetic outbursts from the young Sun, condensation from a cooling gas of solar composition, formation by current sheets in the protoplanetary disc, and plenty more besides [14,18,19]. Scientists always strive to find a simple and elegant explanation for natural processes. But in the case of chondrules, it remains possible, and perhaps even likely, that they were formed in a variety of different ways, not just one.

And so back to Mr Pettifor. He eventually donated his stone to the Natural History Museum for a modest fee in view of the rarity of UK falls. I of course was left with the impression that meteorites falling into people's gardens was going to be a regular part of my life. It seemed clear to me at the time that there would be a steady stream of calls to the Museum, as lumps of rock from outer space rained down at regular intervals on the lovely, manicured lawns of England. And for my part, I would be staying fit, strolling the quiet English countryside doing my bit, searching for extra chunks from each of these meteorite falls. How misguided and deluded I was. In fact, we had to wait another 30 years before a space rock was recovered again on UK soil. We will get to that event in due course.

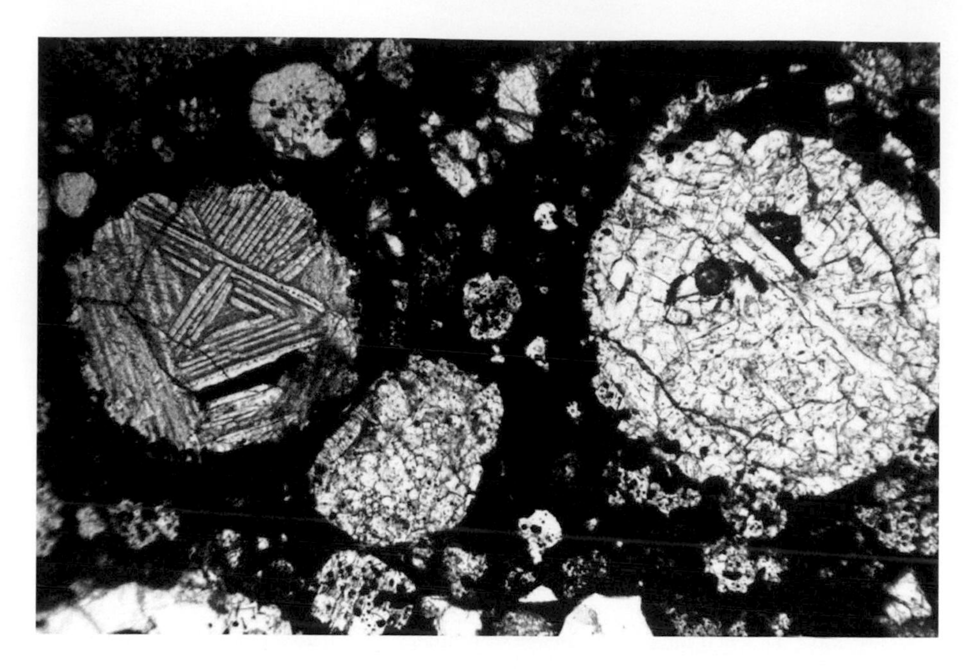

FIGURE 2.7 Chondrules are not just found in ordinary chondrites. This is a microscope view of the carbonaceous chondrite Dar al Gani 067 recovered from the Al Jufrah region of Libya in 1995. Chondrules have internal structures that indicate they were formed by rapid heating followed by rapid cooling. All sorts of processes have been put forward to explain how they formed (see the main text). This particular type of meteorite is from a group known as the CO3s (see Appendix 2), which typically have very small chondrules. The big one on the right has a diameter of about one-tenth of a millimetre. (Image: the author.)

Well, that was a very nice story. Gentleman in garden almost hit by meteorite – narrow escape. All's well that ends well. A very rare event with a nice happy ending. But truth is always stranger than fiction. In fact, small meteorites are falling in gardens every day of the week. And some will fall in your garden too! In the next chapter, we will be heading over to your property to take a closer look.

3 Yes! You Have Meteorites in Your Garden Too! Lots of Them!

You might think that a meteorite landing in an English country garden would be a very rare event indeed. But you would be wrong. It is happening all the time. In fact, there will be literally thousands of them in your back garden. It's all just a question of size.

Spring is a lovely time of year! Following a cold, drab and dreary winter, it is always an uplifting moment when you first catch sight of those early spring flowers – crocuses, snowdrops, daffodils, and bluebells, heralding the onset of warmer, sunnier weather. Days grow longer and brighter. But with more sunshine streaming through the windows, you can't help but notice that layer of dust that has accumulated on various surfaces around the house. Time for a little bit of spring cleaning perhaps? You run a finger gently across the surface of an old bookcase and a thin dusty layer is disturbed. Yikes! Where did all that stuff come from? Household dust turns out to be a pretty complex mix of ingredients, including dead skin cells, hair and textile fibres, dust mites, body parts from dead insects, soil from the garden, microplastics, and pollen grains [1]. A very minor component within that dust will be particles from space [2]. There will not be many and it would be an extremely tough job to isolate them from all those nasties, but yes, you will almost certainly have some very tiny meteorites hanging out somewhere in your home (Figure 3.1).

Estimating how much extraterrestrial material arrives on Earth each year is not easy, but it is likely to be about 60,000 metric tons [3]. Fine-grained space dust is the largest proportion of this total, being about 40,000 metric tons per year [3]. These estimates are subject to very large uncertainties. Love and Brownlee (1999) suggest that the space dust figure could be as low as 20,000 or as high as 60,000 metric tons per year [3]. But taking the numbers at face value suggests that, while the big spectacular meteorites get all the attention, the vast bulk of the space materials arriving on Earth are in the form of very tiny particles. How small? The most important size fraction amongst these grains has a diameter of 0.2 mm and weighs in at a

DOI: 10.1201/9781003174868-3

FIGURE 3.1 From time to time, your trusty vacuum cleaner will contain at least a few extraterrestrial particles. Unfortunately, it will also contain a lot of other stuff that didn't come from outer space. Finding the cosmic dust amongst all the rest of the detritus is the real problem. (Photo: the author.)

minuscule 0.000015 g [3]. But while that sounds tiny, it is still clearly visible with the naked eye, which is able to resolve objects down to at least 0.04 mm in diameter, the width of a human hair [4].

However, that figure of 40,000 metric tons a year is just the amount that hits the top of the atmosphere. Unfortunately, a lot of space dust burns up on entry and so never makes it to the surface. Martin Suttle and Luigi Folco estimate that the annual amount of micrometeorites deposited on Earth is actually in the range of 1,500–6,500 metric tons [3]. If we use their upper estimate and a value of 0.2 mm for the average diameter of cosmic dust, we can make the rough calculation that each year there will be about one of these grains added for every square metre of the Earth's surface [3]. Of course, particle grain sizes vary enormously [3], so that figure is going to be pretty inaccurate. However, it does clearly demonstrate that far from being a rarity that you only encounter in museums, meteorites are everywhere! (Figure 3.2).

But let's stay with this space rocks in gardens thing for a bit longer. Naturally, the size of gardens varies a lot, but the UK Office of National Statistics has very kindly calculated the median garden size in Great Britain, which is 188 m^2 [5]. With 1 grain per square metre that means that the average UK garden receives several hundred space grains each year. That would be a nice haul if you could collect them all.

FIGURE 3.2 On average, one extraterrestrial dust particle lands each year for every square metre of the Earth's surface [3]. That means your garden will contain hundreds, or even thousands of them. Of course, it will contain a lot of other things too! Finding space dust in urban and suburban environments is not easy, but it has been done successfully, as we shall see. (Photo: the author.)

And of course, people don't just own a garden for one year. It is estimated that the average homeowner in the UK retains their property for 23 years [6]. In that time, the average garden would have accumulated well over four thousand cosmic dust particles, which is a lot. And don't worry, cosmic dust is robust stuff and would degrade little on that timescale. From these, albeit rough and ready calculations, you can see that meteorites falling into well-manicured gardens, like that of Mr Arthur Pettifor, is not an exceptional event at all, but the norm. Important to point out that tiny meteorites are not horticultural snobs and will also fall into messy gardens too!

While it is certainly the case that cosmic dust is present everywhere on the Earth's surface, locating individual particles is not going to be easy. A needle in a haystack doesn't get anywhere close. But that hasn't stopped scientists trying. And they have been at it for a long time. Between 1872 and 1876, the Royal Navy vessel HMS Challenger undertook a pioneering survey of the world's deep ocean basins (Figure 3.3). It was a highly successful expedition and made many important discoveries [7]. From our perspective, the most important finding was the identification of small, generally less than 1 mm, rounded, magnetic particles, now referred to as "cosmic spherules". To the great credit of the scientists on board HMS Challenger, they immediately recognised these objects as being particles from space [8,9].

FIGURE 3.3 HMS Challenger in Antarctic ice. Drawing by Royal Navy sub-lieutenant navigator Swire. Dated 16 February 1874. (Image: State Library Victoria [7].)

Another way to collect particles is in the air. Since 1981, NASA has been collecting dust high in the atmosphere using aircraft specially modified for the job [10]. In particular, once the aircraft is at a sufficiently high enough altitude to avoid potential terrestrial contaminants, clean collector plates are deployed. These are coated with a layer of silicon-based oil which traps the particles. Once on the ground, the silicon oil is removed and the particles are imaged and catalogued. NASA maintains a searchable online database of all the cosmic dust particles it has collected and scientists everywhere are free to take a look at what's available and request samples for further analysis [11] (Figure 3.4).

And NASA has a lot of space particles to choose from. In fact, it curates over 11,000 individual extraterrestrial dust particles [11]. There are two main types available. Dust particles collected by high flying aircraft tend to be smaller than about 0.1 mm in diameter and have a fluffy, porous texture. These tiny grains are known as Interplanetary Dust Particles, or IDPs for short (Figure 3.5a), and because of their small size, they have tended to escape significant heating during deceleration in the Earth's atmosphere. Particles larger than about 0.1 mm, but less than 1 mm in diameter, are generally referred to as micrometeorites. These do not tend to be collected directly in the atmosphere, but from surface deposits. In the case of the NASA collection, these larger-sized particles have been retrieved from ice melted at the South Pole [12]. Due to the heat generated as they enter the atmosphere, these larger particles are often highly cooked and may even be melted, forming the rounded cosmic spherules of the type originally located by HMS Challenger. However, depending on how fast they entered the atmosphere and at what angle, some of these larger particles can escape significant heating [13].

You don't actually need to call NASA to get hold of particles from space. After all, there are loads of them in your garden, as we have seen. But as mentioned earlier, the real problem is finding them. During the middle years of the twentieth century, collecting space dust in urban areas had been a popular activity [13–15]. But it was eventually realised that almost all the particles being collected were the byproducts

FIGURE 3.4 A NASA WB-57 aircraft used to collect interplanetary dust particles (IDPs) in the Earth's atmosphere. NASA's high-altitude aircraft collect these particles by deploying collectors coated in a silicon-based oil. These are able to catch the IDPs before they descend to lower altitudes where terrestrial contaminants abound. (Image: NASA.)

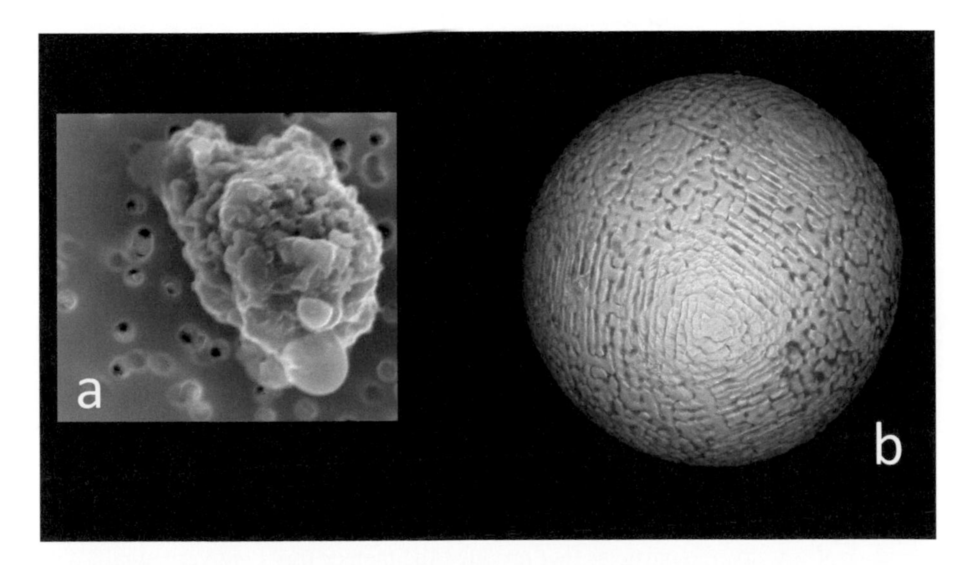

FIGURE 3.5 (a) The particle on the left is an IDP collected in the atmosphere by a NASA high-altitude aircraft (particle U2153-B-3,0). (Image: NASA.) (b) The particle on the right is a cosmic spherule collected from the Transantarctic Mountains. The cosmic spherule has a diameter of about 0.13 mm, which is more than 25 times greater than the IDP shown on the left. (Image: Luigi Folco.)

of human activity, produced by a variety of industrial processes [15]. As a result, the hunt for urban micrometeorites fell out of favour and was generally thought of as a scientifically discredited activity [13]. But that changed in 2009 when a Norwegian jazz musician and geologist Jon Larsen was sitting outdoors one day having his breakfast [16]. He was about to tuck into a lovely bowl of strawberries when he noticed a tiny speck of dust on his table. Nothing unusual there of course. But Jon had only just cleaned it, and this speck of dust was hard and stony. Jon wondered what it was and how it had got there. So, he did a bit of research and discovered the world of micrometeorites [16]. He read about cosmic spherules and their discovery by the scientists on board HMS Challenger. That original piece of dust lying next to a delicious bowl of strawberries may not have been a micrometeorite (of course it could have been), but it set Jon off on a six-year odyssey to identify and collect genuine urban micrometeorites. He collected samples of dust all over the place, on roads, rooftops, gutters, and pavements, in the many countries that he visited as part of his job as a jazz musician. He contacted numerous scientists about his hobby and was almost always told not to bother. The received wisdom was that you can't find micrometeorites in urban areas because they are just swamped by all the other stuff that is out there.

But Jon didn't give up. And then in 2015, he found a particularly unusual looking particle and sought the assistance of Dr Matthew Genge, a micrometeorite expert from Imperial College, London. Matt Genge knew the genuine thing when he saw it and was able to confirm that Jon had indeed located the first fully authenticated urban micrometeorite. But one particle in six years of searching! Was it time for Jon to quit while he was ahead? Of course not! Now he knew what to look for there was no stopping him. In the years following that initial urban micrometeorite discovery, Jon and other groups inspired by his work have located thousands more urban cosmic particles [13,15–17]. There is even a searchable online database [18].

But hunting on rooftops and gutters to locate cosmic particles is not for everyone. No need to worry, you can experience the wonder of space rocks from your back garden in other ways. Looking up at a clear, moonless night sky, away from sources of urban light pollution, it is commonplace to see anything from one or two, up to as many as sixteen, sporadic shooting stars per hour [19]. They are a beautiful sight and often invoke a gasp of wonder from the enthralled observer. Things get even better during meteor showers. These are regular events that take place as the Earth ploughs through a zone of dust and debris left in the wake of a passing comet or, in the case of the Geminid Shower (late December), an asteroid. During meteor showers, the number of events can exceed 100 per hour [19]. But what sort of particles produce these natural firework displays? A case in point is the Leonid Shower. It usually peaks in mid-November and is caused by the Earth passing through the trail left by the comet 55P/Tempel-Tuttle [20]. But while there may be lots of shooting stars during the peak of activity, the particles that produce the meteor trails are very small, with an estimated size of between 1 mm and 1 cm. The principal reason why the meteors are so bright is not their size, but their speed. The Leonid particles are travelling at around 70 km per second when they slam into Earth's atmosphere. The small size of the particles that form meteors and their high speeds mean that they burn up high in the atmosphere at heights of approximately 80–90 km (Figure 3.6).

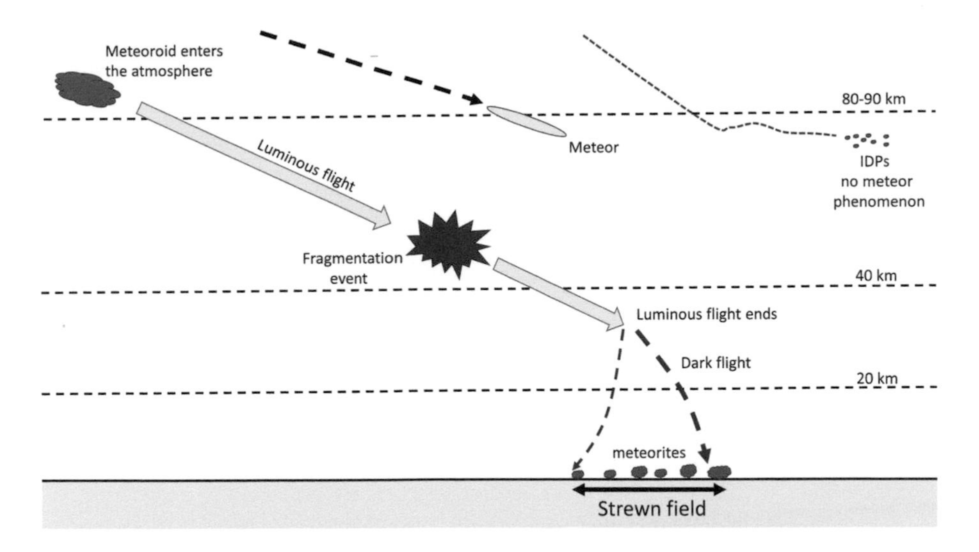

FIGURE 3.6 Compared to larger meteorites, the particles that form meteors tend to burn up high in the atmosphere. In contrast, IDPs, due mainly to their very small size, show none of the phenomena associated with meteors. When the object entering the atmosphere is big enough, it can make it all the way down to the Earth's surface. But it's not an easy ride. The meteorite will experience significant heating, producing a fireball and at deeper levels in the atmosphere, will undergo explosive fragmentation. Once the fireball is extinguished, a period of non-luminous or dark flight takes place before fragments reach the ground to form a strewn field. Note that during dark flight, bigger fragments tend to travel further along the fireball trajectory, whereas smaller fragments are often significantly perturbed by winds in the lower atmosphere. (Image: the author, modified and redrawn from Moilanen et al. (2021) [21].)

Far from being remote objects of little consequence to our daily lives, you can see that space rocks are everywhere. As you walk around, you are continually treading on them (ouch!). They are in your house, in your garden, and they can even fall on your breakfast table as you enjoy a spot of al fresco dining. Look up on a beautiful dark evening and you may catch a fleeting glimpse of them glowing incandescently as they shoot across the sky. And in the later sections of this book, you will learn even more profound ways in which meteorites have shaped our lives.

But first, it's time to find out a little more about the potential dangers posed by falling space rocks. Not the small dusty types we have been looking at in this chapter. No, these ones are much, much bigger and potentially a lot deadlier. Earlier, I said not to get your insurance policy changed. There is a good reason for that. If one of these things ever hits Earth, your insurance company will be vaporised along with the rest of the planet's surface. We are heading to the world of the dinosaur killing asteroids. Hold on to your hat!

4 Dinosaur Killers

Now it's time for all that big dramatic stuff. Giant space rocks causing mass extinctions, etc. But is it true? Were the dinosaurs really wiped out by an asteroid? And what are the chances that it could happen again, and this time it might be us? We are entering classic horror movie territory.

These days, we take it for granted that the poor old dinosaurs were wiped out by a killer asteroid [1] (Figure 4.1). No one really seems to question the idea much anymore. Death by massive volcanic eruption has been a popular alternative, it has to be said, but comes in a distant second place, certainly in the popular imagination [2,3]. A quick perusal of the web and you get the strong message that death by asteroid is the only serious contender for the mass extinction event that took place 66 million years ago [4]. However, it is important to remain cautious. There is detailed fossil evidence that the dinosaurs were already in significant decline before the asteroid arrived [5]. Catastrophe seldom has a single cause.

The asteroid extinction hypothesis is actually a fairly late addition to the list of proposals put forward to explain why the dinosaurs vanished. When I was an undergraduate student studying geology, sometime ago now it has to be said, the abrupt end to the dinosaur era was already a well-established fact. Lots of explanations were put forward to account for this event, with one notable omission. An exploding space rock was never really discussed as a serious possibility. There was the suggestion that dinosaur eggs had been eaten by small mammals and other animals [6]. The dinosaurs perhaps just grew too big, too fast [7], or were brought down by a raft of chronic ailments [8]. One theory that I particularly liked claimed that the dinosaurs suffered from a bout of "eggshell thinning" (Figure 4.2) [9]. This was the idea that a hormone imbalance brought on by climate stress reduced the thickness of dinosaur eggs, resulting in shell breakage and consequent embryo dehydration [9].

Poisoning by the naturally occurring element selenium has also been given as another possible explanation for the end of the dinosaurs [10]. Or perhaps as the climate cooled, dinosaurs, which were generally quite large creatures, couldn't sleep through the cold winters due to a lack of suitable hibernation sites [11]. Death by insomnia! You couldn't make it up, but they did! And there are lots more slightly dotty ideas to go along with the ones already mentioned [7].

DOI: 10.1201/9781003174868-4

FIGURE 4.1 T-Rex as it might have looked. Oxford Museum of Natural History. (Photo: the author.)

When not invoking these slightly bizarre theories, many palaeontologists looked towards climatic change as being the most likely explanation for the end of the dinosaur era. The fact that many other important groups of animals and plants became extinct at this time supported a global event [12]. The end of the dinosaurs, it was argued, needed to be seen in a much wider context (Figure 4.3). The event that occurred at the Cretaceous-Tertiary (K-T) boundary (also referred to as the Cretaceous-Palaeogene K-Pg boundary) [13] is an example of a mass extinction, but it was not the only one [14]. Life on Earth has been punctuated by a number of such periods of catastrophic loss of biodiversity. To paraphrase the British palaeontologist Derek Agar: life on Earth has been defined by long periods of boredom and short periods of terror [15].

But our view of the K-T mass extinction changed dramatically in 1980 when the Nobel Prize-winning physicist Luis Alvarez, his geologist son Walter Alvarez, together with Frank Asaro and Helen Michel of the Lawrence Berkeley Laboratory, provided evidence from sequences of sedimentary rocks in Italy, Denmark, and New

FIGURE 4.2 Dinosaur extinction due to "eggshell thinning" [9] is one of many theories put forward to explain the mass extinction that took place at the K-T boundary 66 million years ago. It is fair to say that asteroid impact is now the most popular idea to explain the demise of the dinosaurs. (Photo: the author.)

Zealand that a giant asteroid was to blame [16]. A thin layer of rock along the K-T boundary was enriched in the precious metal iridium (Ir) by 30–160 times its normal level in the Earth's crust. Iridium is only normally found in very low concentrations in terrestrial rocks, as it was lost to the Earth's metallic core during the early stages of our planet's growth. But iridium levels are high in many types of asteroids. It was a classic case of a smoking gun. The high iridium in the K-T boundary layer sediments could only have had one source and it wasn't from this planet.

The Alvarez hypothesis, as it is often called, proposed that a giant asteroid hit the Earth causing massive climate change (Figure 4.4). In the original 1980 paper, the size of the asteroid was estimated to have had a diameter of 10 km. To put it into perspective, Mount Everest is only 8.8 km high. As the Alvarez team pointed out in their paper, the impact of such an enormous asteroid would have resulted in a global winter from which few species could have escaped.

FIGURE 4.3 Dinosaurs on display at the Houston Museum of Natural Science. The dinosaurs were the most celebrated victims of the huge asteroid that struck the Earth about 66 million years ago. But they were not alone. About 75% of life on Earth was destroyed in the cataclysm. (Photo: the author.)

FIGURE 4.4 The Alavrez hypothesis explains the mass extinction that took place on Earth 66 million years ago as being the result of a giant asteroid impact. It has been estimated that this body was larger than Mount Everest. (Images: NASA.)

When the idea first started to gain traction in the early 1990s, I had a brief conversation with a passing palaeontologist on the Cromwell Road, just outside the Natural History Museum, as one does. I asked him what he thought of the killer asteroid hypothesis. He looked at me as if I had said something very rude and just replied that it was total nonsense. He may have been less polite than that. At the time, a lot of scientists were unconvinced. They thought that it was either an unrelated event or, at best, only a minor contributory factor.

In the years that have followed its initial proposal, the evidence in favour of the theory has grown significantly stronger. The iridium layer has been identified at many locations throughout the world. Furthermore, the location of the impact site itself has now been identified by remote sensing techniques. Multiple lines of evidence point to a large, buried crater-like structure centred on the Yucatan peninsula in Mexico as the location where the asteroid crunched into the Earth [17]. It is now generally referred to as the Chicxulub crater.

But the asteroid impact event that killed the dinosaurs and many other creatures was not an isolated occurrence. And the frequency of such collisions increases as you go back in time. You just have to look up at the Moon and study its scarred surface to realise that huge crater-forming events were more commonplace in the past (Figure 4.5). And what went for the Moon went for the Earth and the rest of the inner Solar System. It is the reason why the cratered surface of Mercury looks so much like our Moon. Shortly after the planets in the inner part of the Solar System reached roughly their present size and orbital configuration, a surge in asteroid impact activity seems to have taken place. An event that is generally referred to as the "Late Heavy Bombardment" [18]. As we will discuss in a later chapter, our Solar System

FIGURE 4.5 The heavily cratered face of the Moon is testimony to the high rate of asteroid bombardment that took place in the inner Solar System early in its history. This period is often referred to as the "Late Heavy Bombardment". It was only "late" with reference to the formation of the inner Solar System bodies, including the planets and the Moon. It actually took place about 4,000 million years ago. (Image: NASA/Lick Observatory/ESA/Hubble.)

likely formed 4,567 million years ago. During the first 100 million years or so, the planets formed and established themselves in their present orbits. Then, for reasons that are still not fully understood, about 4,000 million years ago, all hell broke loose, as the inner Solar System was bombarded by a stream of huge asteroids. The Late Heavy Bombardment lasted until about 3,800 million years ago. Most of the really large craters on the Moon date from this time. This event would also have affected the Earth. Craters as large as those on the Moon, or perhaps even larger, would also have formed on the surface of our planet. But very little evidence of them exists today. Unlike the Moon, the Earth has experienced extremely active erosion by wind and water.

It also has to be said that the Late Heavy Bombardment is not a universally accepted idea and one alternative possibility is that the spike in cratering that seems to have taken place at that time may really have been just the tail end of planetary formation [18,19]. The large inner Solar System planets like the Earth grew by sweeping up all the smaller asteroids in their paths. This process would naturally have started to tail off as less and less material was left over to be incorporated into these growing planets. At first, when collisions were frequent, each new impact would, to a certain extent, have overprinted any earlier ones and reset the record. But as the impact rate declined, the age we now recognise as being that of the Late Heavy Bombardment may just reflect the age of the oldest structures that have left some sort of surface record. Whether or not the Late Heavy Bombardment was a real event, it remains clear that the inner Solar System was a very dangerous place in the early days following its formation. It would have been a precarious place for life to have established itself.

While big asteroids hitting the Earth became less frequent after the Late Heavy Bombardment, they certainly didn't stop. And the Earth's surface still bears the scars of these catastrophic events.

The largest known astrobleme, as these giant craters are sometimes called, is the 2,023-million-year-old Vredefort impact structure (Figure 4.6). It is located in South Africa and is estimated to have had an original diameter of between 180 and 300 km [20]. It was probably formed by an impacting asteroid with a diameter of at least 10 km [20]. The crater has suffered major erosion and what remains now is the deformed underlying bedrock. Like most large craters throughout the Solar System, it has a central uplifted area, which in the case of Vredefort is known as the Vredefort Dome.

The second largest identified crater on Earth is the dinosaur-killing, Chicxulub structure in Mexico, which may have had an original diameter of about 180 km [17] (Figure 4.7).

Based on its current maximum diameter of 130 km, third place goes to the 1,850-million-year-old Sudbury Structure in northern Ontario, Canada [21]. While the Sudbury Structure is well-exposed, it is also highly deformed and doesn't really have a nice, rounded shape typical of more recent craters [22]. Estimates of its original size are difficult to make, but it may have been 200 km in diameter [22]. It has been suggested that Sudbury was formed by a comet rather than an asteroid [23].

The Sudbury Structure was in fact my own first encounter with things extraterrestrial. As a new postgraduate student in Ontario, I was keen to learn about the economic geology of the relatively nearby Canadian Shield. With that in mind, I joined an undergraduate field trip led by the friendly and ever enthusiastic professor

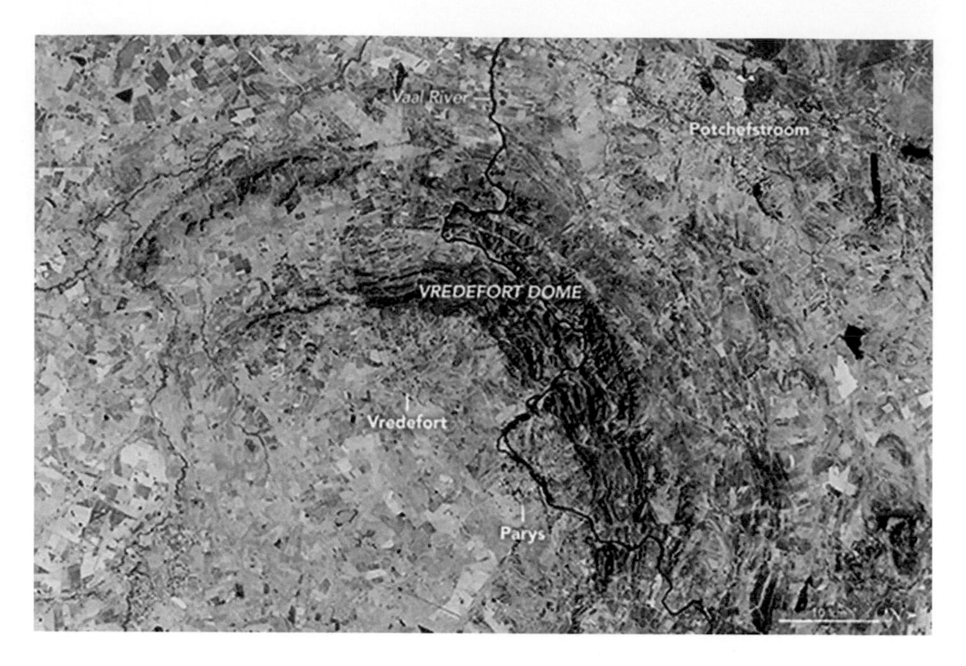

FIGURE 4.6 The 2,023-million-year-old Vredefort impact structure is a large, but highly eroded, crater located in South Africa. (Image: NASA.)

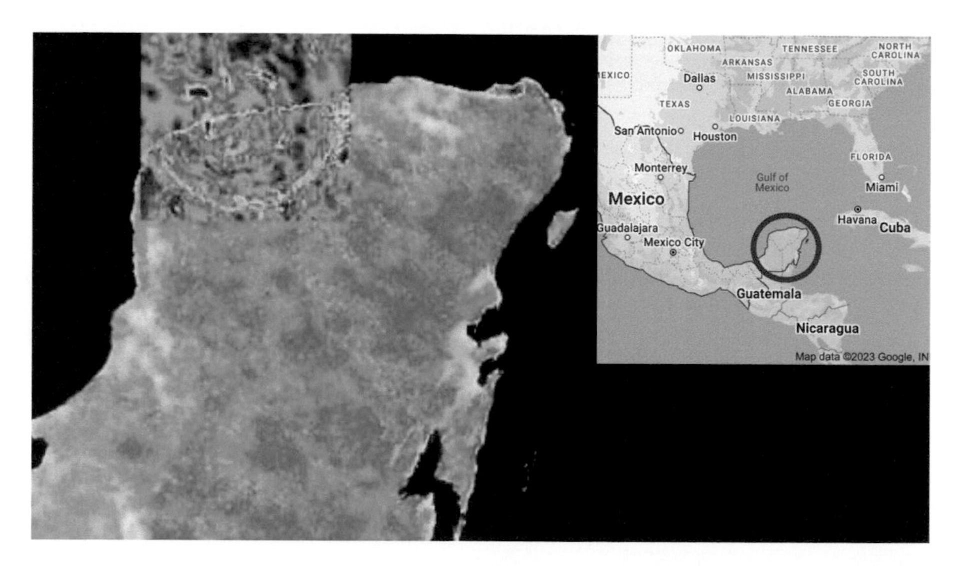

FIGURE 4.7 Gravity map showing the multiple ring structure of the Chicxulub Crater located at the northwest tip of Mexico's Yucatan Peninsula. The crater is now totally buried but is estimated to have had an original diameter of 180 km and to have been 20 km deep. Chicxulub is considered to be the most likely impact site of the K-T boundary, dinosaur-killing asteroid [17]. (Image: gravity map: NASA, location map: Google.)

of economic geology, Richard Hutchinson. We visited many of the main mining camps in Northern Ontario and Quebec, including Noranda, Timmins, and Elliot Lake. We went down a number of important mines, including the spectacular Kidd Creek Mine, now the deepest in the world and, at 2,735 m below sea level, the deepest accessible non-marine point on Earth [24].

One day on the field trip, we had stopped at a roadside cut in the Sudbury area that had been blasted through the local granitic rocks. We were examining strange cone-shaped features, some of which were metres long. These are known as shatter cones [25]. Richard Hutchinson took the opportunity to explain some of the controversy surrounding the formation of the Sudbury Structure. It was a squashed basin, shortened in a north-south direction during large-scale movement of the Earth's crustal layers. Within the basin was a shallow-dipping igneous body, known locally as "the Nickel Irruptive", which along with nearby local rocks, was host to economically important metal deposits. The rocks surrounding Sudbury were smashed and deformed. The shatter cones that we were studying required enormous instantaneous pressures for their formation [26]. In short, the origin of the Sudbury Structure was something of a puzzle.

But many years before that field trip, one man had been in no doubt about how the Sudbury Structure had formed. In the early 1960s, Robert Dietz (Figure 4.8), a pioneer in the study of plate tectonics, visited Sudbury and interpreted it as an asteroid impact structure [27]. He further suggested that the nickel irruptive was the result of

FIGURE 4.8 Robert S. Dietz was the first to propose that the Sudbury Structure was an impact basin. (Image: Scripps Institution of Oceanography.)

surface volcanic activity similar to that which had produced volcanic rocks on the Moon, known as Lunar Mare basalts. Based on Richard Hutchinson's comments, it appeared that this had not gone down well with the local geologists and there had been a lot of resistance to this space age explanation. We had a bit of a discussion on the outcrop, as one does on such field trips. Someone asked whether the shatter cones might have been formed recently by Highways Canada when they had come through blasting the local rocks during road construction? But that didn't seem likely as the shatter cones were everywhere in the Sudbury area, most being a long distance from the best efforts of Highways Canada. Then there were the PDFs in quartz, standing for Planar Deformation Features [28]. PDFs are basically microcracks in quartz crystals. Again, these could only have formed during a process that involved very high instantaneous pressures. To be honest, based on what I had seen on the field trip, the impact model for Sudbury seemed the most likely explanation to me, and to most of my fellow students as well.

But it soon became clear that even amongst academics not everyone liked the impact model. Back at the university, in a lecture given only a few weeks after the field trip, one of the other professors presented a non-impact model for Sudbury. He really didn't like the asteroid impact model one little bit. There was a very lively and not very polite discussion at the end of his presentation. Today, no one seriously doubts that the Sudbury Structure formed as the result of a giant impact event. If life had been a bit more developed back then, rather than just a bunch of unicellular organisms, there can be little doubt that the effects from the Sudbury impact would have been just as catastrophic as the later dinosaur-killing event.

Well, so much for the past! What about the dangers from giant impacts today? In Chapter 1, we saw that no one has yet been wiped out by a meteorite. In fact, if you want to worry about hazardous events, there are plenty with more down-to-Earth origins.

In terms of death rates from natural disasters, flying space rocks are absolutely at the bottom of the list. Earthquakes are by far the deadliest events with approximately 24,000 fatalities annually, extreme temperatures account for between 7,400 and 17,000 deaths per year, followed by floods with about 5,000 victims and storms with about 2,000 [29]. Deaths from volcanic eruptions tend to be very episodic, with estimated annual numbers for the recent past in the range 28–136 [29], but this might increase to 540 victims a year if a timespan of centuries is considered [30]. Skin cancer, at least in part caused by exposure to sunlight, is estimated to result in about 60,000 deaths per year worldwide [31]. If we look at more specific natural phenomenon, such as being struck by lightning, this has an estimated annual death rate of 4,101 [32], which does seem extremely high, but you get the general idea. Compared with no known victims, almost every natural event is potentially more dangerous than the chances of being killed by a meteorite. Clearly, as we have already discussed, there would seem to be no pressing need to change your insurance policy to cover falling space rocks.

But here is the thing. The fact that no one has been killed in recent times is not a very good predictor of the potential threat from asteroids. A famous historic example of a natural disaster illustrates this point nicely. In the years leading up to the catastrophic AD 79 eruption of Vesuvius, there would have been few indications of the potential disaster that awaited the inhabitants of the cities of Pompeii and

Herculaneum. Then in a single event lasting just 15 minutes, it is estimated that 2,000 people were asphyxiated by hot gases and ash [33]. An episodic catastrophe may give little warning but is deadly nonetheless. And there have been some relatively recent events that suggest meteorite impacts have the potential to be extremely hazardous. Perhaps the most relevant is an incident that took place in Russia in 1908.

On the morning of 30 June, a colossal explosion took place over a remote region of Siberia, known as Tunguska. It is estimated that an area of forest exceeding 2,000 km^2 was devastated in the blast, with some 80 million trees being destroyed as a result. Prior to the explosion a bright fireball was witnessed over a wide area and this was followed almost immediately by a series of flashes and sometime later, depending on the distance of the observer from the epicentre, a shock wave was experienced. Eyewitness reports collected many years later indicate that people were thrown to the ground by the energy from the blast [34].

From a scientific perspective, one of the unfortunate aspects of the event was the long delay before an organised expedition reached the blast zone. It wasn't until 1927 that a serious attempt was made to study Tunguska, although there had been a number of earlier expeditions to the area. It was expected that a crater of some type would have been formed, but this was not the case. Instead, a large area of felled trees was discovered [35] (Figure 4.9). At the centre was an 8 km wide area of upright, but denuded trees. Away from this central zone, the trees had all been felled and lay on the ground pointing radially away from the centre of the blast zone.

Due to the high level of uncertainty concerning the nature of the Tunguska "event", a wide range of diverse scenarios have been put forward to explain what took place. It has been suggested that the projectile was a stony meteorite, an icy comet, a

FIGURE 4.9 Fallen trees in the Tunguska area. (Photograph from The Soviet Academy of Science 1927 expedition led by Leonid Kulik.)

natural nuclear bomb, the result of antimatter and/or a black hole. Even visiting aliens have been invoked [35]. The suggestion that the Tunguska meteoroid was a cometary body, and hence composed of ice and dust, was first proposed by the British scientist Francis Whipple in 1930 [35]. It has remained a popular explanation for the event ever since. The size of the Tunguska body is unknown, but some estimates have put it at 50–60 m in diameter. There is a general consensus that the lack of a central crater and the distinctive orientation of the felled trees points to an air burst explosion. This would have taken place when the incoming projectile penetrated into the denser lower atmosphere with the consequent build-up of forces at the front of the body. Once these forces had exceeded the intrinsic strength of the projectile, it would have violently disintegrated.

It is unfortunate that the Tunguska event is so poorly understood as it has important implications for our understanding of the threat posed by large asteroids or comets. There can be little doubt that if the Tunguska explosion had taken place over a major population centre, the loss of life would have been extremely large. Tunguska serves to demonstrate the real and present danger posed by space rocks. The events at Tunguska were an important motivation for NASA to undertake the very successful Double Asteroid Redirection Test (DART) mission [36]. But more on that a little later.

While some potential extraterrestrial material has been recovered from the Tunguska area, it has unfortunately been of either poor quality or ambiguous composition. The analyses that have been undertaken have failed to settle the issue surrounding the identity of the projectile. However, this paucity of recovered material appeared to have come to an end when a large sandstone boulder called John's Stone was identified as being debris from the Tunguska explosion [37]. But sandstones are also very common rocks on Earth and unknown amongst the meteorite population so far studied. The team involved with John's Stone suggested it might be a new type of meteorite, possibly from Mars. Now this was very exciting. Mars is known to have layered rocks that most likely formed in shallow lakes and seas on the ancient surface of Mars [38]. And we now have a large bunch of meteorites from Mars. In fact, the Meteoritical Bulletin database currently (December 2023) lists 366 distinct samples that have a well-documented origin from the red planet [39]. But here is the curious thing. Despite all the evidence from Mars for sedimentary rocks on its surface, not one of the so far identified Martian meteorites is of exclusively sedimentary origin. It's a mystery. We should have them, but they have yet to be identified. But perhaps John's Stone would change all that.

Following publication of these claims, I was contacted by Henning Haack of the Natural History Museum of Denmark, who had studied the Tunguska event in some detail. He naturally felt that these extraordinary claims needed to be followed up and verified, or otherwise, as soon as possible. If John's Stone was from Mars, there was one test that could prove it beyond reasonable doubt. Martian meteorites have a very characteristic oxygen isotope composition [40]. That's where I came in. If John's Stone was from Mars, our analyses would provide conclusive proof.

Dr Yana Anfinogenova, a member of the Russian team, was extremely helpful and supportive of our new study, providing us with samples of John's Stone. Henning also contacted Henner Busemann of ETH Zurich, a specialist in noble gas analysis of extraterrestrial samples. Unfortunately, our combined oxygen isotope and noble gas analyses were conclusive, John's Stone was not a space rock but a relatively normal terrestrial sandstone [41]. The results of our study were published by the journal Icarus, where the original John's Stone paper had appeared. A reply to our findings was posted elsewhere by the original authors [42].

It was disappointing that John's Stone did not prove to be debris from the Tunguska explosion. And so, the hunt for the real culprit continues. Is this all just of academic interest? Well no, not really. As we have seen, had the Tunguska event taken place over a population centre, the results would have been appalling. What happened at Tunguska needs to be studied and understood. The Russian team that published the original John's Stone paper might not have correctly identified the Tunguska meteorite, but at least they were trying to locate it and one day may succeed. And that would be a very big breakthrough.

But what about potential casualties from the Tunguska event? It was a huge explosion, albeit in a very remote region of Siberia. Recent estimates put the energy of the Tunguska blast in the range of 10–15 megatons of TNT [34,35]. That is 1,000 times more energy than was released from mid-twentieth century era atomic bombs. While the explosion likely took place at a height of 6–12 km, based on the damage caused to the underlying forest, a large amount of energy was deposited at ground level [34]. The people living in the region where the explosion struck were nomadic reindeer herders, known locally as Evenki. A recent study has looked at contemporary eyewitness accounts and estimates that about 30 people were likely present in the areas where the trees were felled by the blast [34]. All these people would have been rendered unconscious by the air burst explosion and have suffered varying degrees of injury. Analysis of these accounts suggests that at least three people may have been killed by the blast. Clearly, this has not been officially certified, but the evidence seems credible that the Tunguska event did result in significant injuries and possibly some fatalities. If this analysis is correct, then meteorite strikes are starting to look a little less benign than we have so far given them credit for.

And now for the cosmic equivalent of that old saying: "you wait ages for a bus and then two come along at once". Back in February 2013 the world was waiting for asteroid (367943) Duende [43]. It was due to pass just 27,700 km from the Earth's surface. To put that in context, the Moon is approximately 384,000 km from Earth. That means Duende was going to be 14 times closer to the Earth than the Moon. That's close. In fact, Duende came so close that its orbit was altered by this terrestrial encounter. Duende has a diameter of 30 m, but at the time of the 2013 flyby it was thought to be somewhat larger, with an estimated diameter of about 45 m. It had been well established in advance that the 2013 close encounter posed little risk to the Earth, but the chances of a nasty event were higher for encounters taking place between 2026 and 2069. Based on its predicted size, an impact by Duende would have been slightly less energetic than Tunguska, but not by much. If it entered the Earth's atmosphere, it would explode with an energy of 2.4 megatons of TNT. In the wrong place that could cause a lot of damage.

But while the world waited for asteroid Duende to do a close, but uneventful flyby, along came a totally unscheduled asteroid, which penetrated deep into the atmosphere and then exploded close to the city of Chelyabinsk (population 1.2 million), in the Southern Urals region of Russia (Figure 4.10). The Chelyabinsk asteroid [44]

FIGURE 4.10 Vapour trail left by the Chelyabinsk asteroid on 15 February 2013. (Photo: ESA/M. Ahmetvaleev.)

arrived 16 hours before Duende and later analysis of its trajectory showed that the two asteroids were unrelated. Their near simultaneous arrival was apparently just a coincidence.

The Chelyabinsk asteroid was slightly smaller than Duende, with an estimated diameter of 20 m. Prior to atmospheric entry Chelyabinsk, unlike Duende, was completely undetected. Chelyabinsk entered the atmosphere at a shallow angle of about 18.5° and was travelling at a speed relative to the Earth of 19 km/s (approximately 43,000 mph). The fireball that formed as Chelyabinsk penetrated deep into the atmosphere was visible up to 100 km away. Finally, at an estimated height of 30 km, it violently disintegrated in an air burst explosion [45]. Most of the asteroid was disrupted into a cloud of tiny fragments only a few centimetres in diameter [46]. These are sometimes called Chelyabinsk "peas" (Figure 4.11). One large metre-sized block did continue along the flight trajectory and finally crashed into the ice-covered surface of nearby Lake Chebarkul and was recovered from the lakebed in October 2013.

The air burst explosion close to the city of Chelyabinsk produced a shock wave that caused some property damage and blew out windows over a wide area. Flying debris, especially broken glass, resulted in injuries to about 1,500 people, luckily there were no fatalities. The energy released by the blast is estimated to have been equivalent to approximately 500 kilotons of TNT. While it was probably less powerful than the Tunguska event, Chelyabinsk demonstrated the potential havoc that a relatively small-sized asteroid could cause if it fell close to a large population centre. Chelyabinsk bought home the fact that while they might be infrequent events, asteroid impacts have the potential to cause a significant level of damage. As a result of Chelyabinsk, authorities around the globe began to take the risk from asteroid strikes more seriously. The events at Chelyabinsk prompted NASA to start work on a new space defence programme [47].

Chelyabinsk was a big media event. The fact that the fireball and subsequent air burst explosion had been captured on numerous car dashboard video cameras (dashcams) was a big help. A good news story needs pictures and Chelyabinsk delivered in spades. Almost instantly, there was a big demand from TV and radio news organisations for scientists to give interviews explaining what had happened.

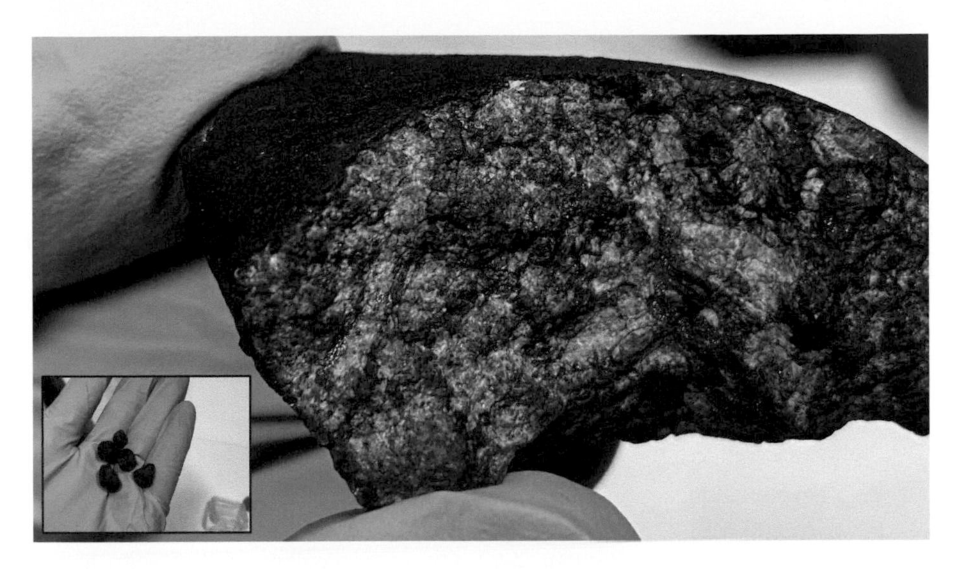

FIGURE 4.11 Pieces of the Chelyabinsk meteorite. Inset photo shows five Chelyabinsk "peas". When the meteorite exploded, much of the material disintegrated into pieces only a few centimetres in diameter. The larger single piece of the meteorite shown here is partially covered in a thin, dark layer of fusion crust. The interior generally has a light colour but is crisscrossed by black shock melt veins. These have nothing to do with the events of 15 February 2013 but were most likely the result of collisions that took place in the asteroid belt. (Image: the author.)

It was all hands on deck, as a sizeable cohort of meteorite researchers tried to rapidly turn themselves into knowledgeable pundits, myself included. As is usually the case, the initial information about what had actually taken place was sketchy. But we did our best.

At the same time as all this media activity was taking place, there was a scientific job to be done. The head of our group, Professor Colin Pillinger, had some high-level contacts amongst the Russian scientific community. He quickly got an agreement for the transfer of some sample material to the UK. But speed was of the essence. We needed to get the precious Chelyabinsk fragments back to our labs for analysis as soon as possible. However, there was no question of anyone flying out to Chelyabinsk, so what could be done? Colin, ever resourceful, worked his contacts. A few days after the fall of the meteorite, we had samples of Chelyabinsk in our labs undergoing analysis. How was it done? Well, let's just say that if you ask them, the UK's diplomatic service will neither confirm nor deny that they had any involvement in the transport and delivery of the Chelyabinsk samples. So, let's leave it at that.

A week or so after the fall of Chelyabinsk, Colin asked me if I wanted to be involved in a documentary about the meteorite that was being filmed by a fast turnaround media company. It sounded like fun, so I said yes. Next, I found myself in a very small studio close to Covent Garden, London working on the film. My job was to sit in front of a video monitor looking at dashcam footage of the fireball and subsequent air burst explosion and then, using a knowledgeable scientific tone, explain what was going on. After several dozen takes, fatigue began to set in. But the producer and cameraman kept up the pressure. Eventually they got the shots they needed. Meanwhile, in Russia, a PhD student from Cardiff and a lone cameraman were getting some relevant material in the field. The whole thing was then spliced together and an interesting narrative angle added which I had been largely unaware of while doing the filming in Covent Garden. The film suggested we were all part of a single team investigating the fall of the meteorite, which was stretching it a bit. It was a long way from the world I was used to, but fun, nonetheless. The film was called Meteor Strike [48] and it appeared in the UK on Channel 4 and was also shown in several other countries. Meanwhile, the BBC was making its own fast turnaround documentary that involved other members of the UK planetary science community. Both films aired at more or less the same time, but had very distinct styles. When a rare event like Chelyabinsk takes place, you have to be willing to drop everything and engage with the media.

Health issues with an extraterrestrial angle briefly made news headlines following a meteorite fall in Peru in 2007. On 15 September of that year, a fireball and ground impact took place close to the village of Carancas, not far from Lake Titicaca [49]. An impact crater about 14 m across was formed and quickly filled with water. Eyewitnesses claimed that the water in the crater began to bubble shortly after the fall event and that a column of smoke rose from it and persisted for a short period. It was claimed that people who came to look at the crater began to fall ill and suffered from a variety of symptoms, including skin rash, headaches, nausea, and diarrhoea. The story was quickly picked up by world's media [50]. The combination of an alien space rock and a mysterious disease proved irresistible. With almost no facts to go on, meteorite experts were sceptical. People falling ill due to strange substances coming from a freshly fallen meteorite had never happened before. In interviews with the media, experts cast doubt on the whole episode. They thought it was unlikely to have been a meteorite fall, perhaps some sort of natural explosion, like a geyser, or that sort of thing. Accidental detonation of abandoned artillery shells was another possibility that was mentioned. But others were less sceptical. The meteorite hunter Mike Farmer visited the site and collected material [49,51]. He encountered a few problems with the local officials and then had to leave relatively abruptly [51]. Scientists in the region also got to work on samples provided by local people and quickly confirmed that the crater-forming object was indeed a meteorite.

What then caused all the health problems associated with the fall of Carancas, so widely reported by the international media? A study published in 2009, while mainly focussed on the formation of the crater, did also look into these health-related issues [52]. The authors are highly sceptical about the claims of medical problems following the arrival of the Carancas meteorite. They interviewed a nurse who worked at the local health centre at the time of the incident. She suggested that there were no reports of adverse symptoms from people who visited the crater in the first few days after the impact. However, that changed when it was announced that a group of doctors and nurses from the regional capital Puno would be visiting to help with any health care issues related to the fall of Carancas. Approximately, 180 people then came forward suffering from a variety of ailments. With reference to the reported medical issues, the study reached some uncompromising conclusions: "*Most of the alleged health problems were a consequence of mass hysteria and people seeking free medical support for pre-existing conditions*" [52]. With apologies to Mark Twain, it would seem that the reports of health issues related to the Carancas meteorite fall had been grossly exaggerated.

Uncertainty of a very different type surrounds the historic Ch'ing-yang event that took place in China in 1490 [53]. Contemporary accounts tell of approximately 10,000 deaths due to rocks falling from the sky. The details are unclear, but could be interpreted as a meteorite airburst explosion similar to Chelyabinsk or Tunguska, but of a much greater magnitude. It seems unlikely that these reports can ever be verified, unless meteoritic material can be recovered from the area and further supporting evidence obtained.

There are a number of other historic accounts of human casualties from falling space rocks [54]. Some reports are more plausible than others, but all have basically little firm scientific evidence to back them up. A Franciscan monk was said to have been killed in Milan, Italy, by a stray meteorite which struck him on the thigh sometime between 1633 and 1664 [53]. Then there was the case of the farmer in Kentucky who died in his sleep after being hit by a meteorite during the night of 14 January 1879 [53]. Two unfortunate sailors are claimed to have met an untimely end on the good ship Malacca during its passage from Holland to Batavia. The culprit is said to have been a 3.5 kg meteorite [53]. There are a number of contemporary documents reporting that one man was killed, and another seriously injured by a meteorite strike in Sulaymaniyah, Iraq in 1888 [54].

And finally, we come to a fully authenticated case of a significant injury caused by a falling meteorite. In November 1954, Mrs Ann Hodges was taking an afternoon snooze under a quilt on her sofa in Sylacauga, Alabama [55]. At 2:46 pm precisely, she was rudely awakened by a 5.6 kg meteorite that crashed through the roof of her house, bounced off the nearby radio and then hit her in the thigh. This resulted in a very large and presumably very painful bruise. She did pay a visit to the local hospital the next day, but this seems to be more to do with all the excitement generated by the event and she seems to have suffered no long-term health consequences. Ownership of the offending meteorite was a bit more complicated to sort out. Mrs Hodges and her husband rented their house and the landlord involved, Mrs Birdie Guy, decided she was the rightful owner of the space rock. A court battle ensued which was settled when the Hodges paid Mrs Guy $500 to keep the meteorite. Eventually, Mr Hodges decided to sell it, but couldn't find a buyer! It did service as a door stop before the Hodges donated it to the Alabama Museum of Natural History [55]. How different it would be today! The Hodges would have had no problem finding a buyer and Mrs Hodges would have been a media star.

While there remains no official evidence linking the death of a named individual to the fall of a meteorite, this does not mean that the risks associated with the unexpected arrival of a biggish piece of extraterrestrial material are zero. As we have seen in the case of the dinosaurs, the fall of a large asteroid can pose an existential risk to life on Earth. Historical events such as Tunguska and Chelyabinsk demonstrate that smaller falls also have the potential to cause catastrophic devastation, albeit on a more local scale. The DART mission [56] (Double Asteroid Redirection Test) which took place on 26 September 2022 is part of the NASA's response to the asteroid threat. DART was a test of the feasibility of altering the orbit of an asteroid by impacting a spacecraft into it. The results were a spectacular success. DART was able to shorten the orbit of the target asteroid Dimorphos by 32 minutes, a figure that greatly exceeded even the most optimistic expectations. There is a long way to go, but the results from DART should allow all of us to sleep a little bit more comfortably in our beds.

So far, we have really been looking at space rocks as potential hazards. And they are. But there is another, more romantic side to meteorites. They are also exotic, mysterious objects with hidden secret pasts that we can only glimpse at. What do I mean by all this? Well, it's time to travel to the hot deserts of Saudi Arabia and meet a space rock that was discovered by a spy.

5 The Spy, the Meteorite, and the Lost Legendary City

Meteorites are certainly epic. But here they take on a truly legendary role as we hunt for a mythical lost city. On the way, we meet up with a very English spy and discover that deception is not just a human quality.

For a short time, I had a job working for a geophysical exploration company based in Dhahran, Saudi Arabia. It was a pretty dull occupation, running data crunching programs in a sort of air-conditioned Nissen hut. I was looking forward to leaving and taking up a place on a geology course in Canada, when news came through that my opposite number in the field had gone berserk and landed a punch on the pilot. This was not his first tantrum and so it was decided to bring him back to town for a spot of rehab. As a result, in the short term, there was a bit of a problem down in the desert and I was dispatched to fill the breach. Where was I heading? Well, the field team at the time were shooting seismic lines in the heart of the Rub' al-Khali desert, otherwise, and with good reason, known as the Empty Quarter. I was not best pleased, my aim back then was to get out of Saudi Arabia, not head off on a jolly to the interior. However, it turned out to be an amazing experience. Although it was a desert in the very strictest sense of that term, it had a strange, haunting beauty. Mountainous red and orange-coloured sand dunes alternating with vast, table-flat sabkhas (Figure 5.1).

This strange terrain created a perfect natural landing strip for the light aircraft that were used to service the field crew. I was only there for a couple of weeks, but it has certainly left a lasting memory. However, and here is the thing, although I didn't realise it at the time, our field camp was only a relatively short distance from one of the sites that has been put forward for the location of legendary lost city of Wabar. Of course, no one really knows the location of Wabar, or even if it ever existed. But that hasn't stopped people trying to find it. And in a sense, whether it ever really did exist may not be the most important thing about this whole story. Perhaps what counts is the fascination and wonder that such places provide. Myths and legends of fabled

DOI: 10.1201/9781003174868-5

FIGURE 5.1 The Rub' al-Khali desert or Empty Quarter of southern Arabia has a strange haunting beauty. Mountainous sand dunes alternate with tablecloth flat sabkhas. Perhaps this is not the ideal site to find an ancient city. And yet legend has it that the mythical lost city of Wabar is located within this dramatic and inhospitable terrain. (Photo: Beda Hofmann, Omani-Swiss meteorite search.)

lost cities motivate and drive unique individuals to battle against great adversity in pursuit of a dream. And the rest of us love to read about their exploits from the cosy comfort of our living rooms. Well, I do anyway!

You may be wondering what all this has to do with meteorites? A legitimate question and I promise we will get to the extraterrestrial stuff very shortly. But first, we need a bit more detail about these ancient lost cities.

Everyone has heard of Atlantis, the fabled island first mentioned by Plato, which disappeared beneath the ocean waves, never to be seen again [1]. But people are much less familiar with "the Atlantis of the Sands", also known as Ubar, Wabar, or Iram of the Pillars, a location mentioned in the sacred Quran [2]. It was T.E. Lawrence, more popularly known as Lawrence of Arabia, who drew parallels between Wabar and Atlantis (Figure 5.2).

But in reality, they don't have that much in common, except perhaps that both came to rather sticky ends, if either actually existed that is. Legend has it that Wabar was destroyed, either as the result of a natural disaster or as a punishment from God. Finding the remains of Wabar has been the quest of a series of celebrated explorers, the most recent being Sir Ranulph Fiennes, who gave an account of his hunt for the legendary site in the 1992 book "Atlantis of the Sands – The search for the lost city of Ubar" [3]. The expedition discussed in Sir Ranulph's book was jointly led by the archaeologist Nicholas Clapp and involved excavations at the Bedouin well at Shisr in the Dhofar province of Oman. As a result of these archaeological studies, it

FIGURE 5.2 T.E. Lawrence, otherwise known as Lawrence of Arabia, was fascinated by the legend of Wabar and called it "the Atlantis of the Sands". The picture shows Lawrence on a camel during the Battle of Aqaba in 1917. (Photo: Imperial War Museum Collection.)

seemed initially that Ubar had been found [4]. But more recent excavations have cast doubt on such claims [5]. While the site at Shisr is relatively ancient, the structures that have been uncovered seem to be on too small a scale to be that of Ubar. David Miller concludes: "Arabia's most famous lost city is still lost" [5].

The Dhofar region of Oman was not the only possibility for the location of Wabar (or Ubar). There were also rumours and stories that it lay to the north amongst the great shifting sands of Saudi Arabia's Rub' al-Khali desert that occupies much of the southern half of the Arabian Peninsula. And one man was determined to try and find it. He was Harry St John Bridger Philby, an explorer, writer, diplomat and spy. In fact, that last "trade" ran in the family, because he was none other than the father of Harold "Kim" Philby, the notorious Cambridge spy. In recent times, the focus of attention has largely been on the damage done to British interests by Philby the younger. The Cambridge spy ring, of which Kim Philby was perhaps the most important member, did untold damage to the UK's national interests during the Cold War period [6,7]. And yet, as hard as it is to believe, Philby the older probably did significantly more damage to his country than his son would later inflict. As a father and son double act in treachery, they have little competition. What then did Philby the older do that was so damaging to UK interests?

Like all great spies, the life of Harry Saint John Philby was highly ambiguous. For every fault, there was a compensating positive attribute. At least that is how Philby would have liked things to appear. Ambiguity and deception are the supreme qualities of a gifted spy, and Philby was one of the best (Figure 5.3).

Philby was born in 1885 and educated at Westminster School and Trinity College, Cambridge. In 1908, he went to India to start a career in the civil service. While in India, he seems to have begun his work as a British intelligence officer. He was transferred to the Middle East during the Arab Revolt against the Ottoman Turks and became an advisor to Ibn Saud, the first King of Saudi Arabia [8,9]. Philby was an Arabist who converted to Islam in 1930. He was widely known under the assumed name of "Shaikh Abdallah". As a close advisor to Ibn Saud, Philby worked against the interests of the British and played a pivotal role in securing the lucrative Saudi Arabian oil concession for ARAMCO (Arabian-American Oil Company) [10,11]. ARAMCO has gone on to become one of the richest companies on the planet and the loss of the Saudi oil concession was a devastating blow to British interests in the Middle East. There were other instances in which Philby appears to have been working to undermine the position of the British in the region and it has been suggested that he bore a grudge against his former employer, possibly as a result of being sacked from his Indian Civil Service job. But Philby was also a man of many compensating talents. He undertook extensive exploration of the Arabian Peninsula, for which he received the Royal Geographical Society's prestigious Gold Medal in 1920. He was also a talented naturalist and an expert linguist, fluent in a wide range of languages,

FIGURE 5.3 Harry St. John Philby, the explorer who found the Wabar craters in 1932. (Image: Wikipedia.)

including Urdu, Punjabi, Baluchi, Persian, and Arabic. He died in 1960 while visiting his son in Beirut. His final words were reported to be "God, I'm bored" [7].

You are probably still wondering what all this has to do with meteorites. Well, not long to wait.

In 1931, Philby was not a happy man [12]. For a long time, he had harboured an ambition to become the first western explorer to cross the Empty Quarter. But then his fellow countryman and former friend, Bertram Thomas, beat him to it, leading a rapid and successful expedition across the Rub' al-Khali starting in Oman. Philby was devastated and felt totally betrayed. For a bit! And then he got over it, as he always did. He had other kites to fly. In particular, there was Wabar. Philby had first heard of it back in 1918 from his Bedouin guide, Jabir ibn Farraj [12]. He recounted the legend that Wabar had been devastated by a wind of fierce destruction that had been sent in retribution for the wickedness of its dissolute ruler 'Ad, who had failed to heed the warnings of his brother Hud [12]. Hud is sometimes identified as the Old Testament prophet Eber [13]. Wabar was said to be located at a place known as "al-Hadida", where close by was a lump of iron the size of a camel's hump. Philby thought that this must be some sort of iron artefact, with the implication that al-Hadida had been a centre of great commercial importance. Hadida means a piece of iron in Arabic.

Finally, in January 1932, Philby got permission from his friend King Ibn Saud to mount an expedition to search for Wabar and duly set off with 15 tribesmen and 32 well-trained Ramliyat camels. They took with them bags of rice, dates, and coffee, sufficient for 75 days of travel [12]. For drinking water, they would make use of local wells, although they were few in number because the Empty Quarter was not a place many local people strayed into. Conditions were extremely tough, such that half the party gave up and returned home. Even with the provisions that they had taken with them, two camels had to be killed for food. But Philby was on a mission and pushed on regardless of the hardships they had to endure. He wanted to be taken seriously as a desert explorer and more than that he wanted to be the person that finally located Wabar. He was assisted by his two Arab guides, Ali Jaham and Salim ibn Suwailim, who about a month into the expedition successfully navigated the group to the Wabar site (Figures 5.4–5.6).

When Philby first gazed on the site, there is no doubt that he was devastated and totally demoralised by what he found, for this was no long lost city. Philby wrote in his journal "*I knew not whether to laugh or cry...*" [12]. What Philby found were two craters which he thought at the time were part of a volcano. He was in no doubt that the legend of Wabar was based on these structures and wrote "*whose black walls stood up gauntly above the encroaching sand like battlements and bastions of some great castles*" [12]. They searched the area for four days but did not find any of the large pieces of iron that legend suggested should be strewn across the site, just a "*silly little fragment of iron about the size of a rabbit*".

The Bedouins were unwilling to accept that the lost city of Wabar was other than had been described in legend. Ali Jahman is said to have disagreed with Philby's view that these were volcanoes and to have told him: "*No, they are the castles of 'Ad. They are his mansions for sure and see how the bricks have been burned with fire, as they relate!*" [12]. The discovery that the basis of the Wabar legend was a natural phenomenon was unexpected and almost certainly not what he had wanted to find, but as usual he made the best of it. Philby was as resourceful and pragmatic as ever.

Following the expedition, Philby took a small sample of the metal fragment he had collected at Wabar to Dr L. J. Spencer of the Natural History Museum in London. It didn't require too much effort to identify the material as being meteoritic in origin. In particular, due to its high nickel content compared to terrestrial rocks.

Since Philby's original discovery, there have been a number of other expeditions to the site, most notably in 1966, when a joint team from the National Geographic Magazine and the ARAMCO oil company were able to locate and remove several large iron meteorite fragments [14]. These are now on display in the National Museum of Saudi Arabia in Riyadh. In 1995, Eugene Shoemaker and Jeff Wynn undertook a study of the geology of the Wabar craters [15,16]. They concluded that there were three distinct craters at the site, an 11 m diameter crater (Figure 5.4), a 64 m crater they called Philby A (Figure 5.4) and a 116 m crater they called Philby B (Figure 5.5). They suggested that the original meteorite, prior to breakup, weighed in at 3,500 tons and arrived from the northwest at a very shallow angle of between 20° and 45°.

In 2008, a Swiss research team led by Edwin Gnos and Beda Hofmann visited the site during a fieldtrip with the Saudi Geological Survey and conducted further field studies [17]. They found that only two craters were visible (11 m diameter crater and Philby B) and that a significant amount of sand movement had taken place since previous recent expeditions to the area. They were able to establish that the sand dunes in the Wabar area migrate at rates of 1 to 2 m per year. Analysis of dark-coloured, impact melted glass fragments collected from the site showed they comprised a mixture of aeolian sand and iron meteorite-derived material (Figure 5.6). Another interesting discovery made by the Swiss team was that the craters do not penetrate below

FIGURE 5.4 Wabar craters visible on the surface taken from adjacent dunes. The small crater on the left has a diameter of 11 m and is known as the "11-Meter" crater. The partially filled crater on the right is the "A" crater discovered by Philby in 1932. Part of a magnetometer used for a magnetic survey conducted on the site can also be seen in the image. (Photo: Jeff Wynn and Eugene Shoemaker/USGS.)

FIGURE 5.5 A view of the 116 m diameter Philby B crater, outlined by the red circle. This image was taken in 2008 by a Swiss team led by Beda Hofmann and Edwin Gnos. (Photo: Beda Hofmann.)

FIGURE 5.6 Small, black, shiny beads (left) and larger black and white "rocky" lumps (right) litter the area of the crater. These are not meteorite fragments but fused desert glass that was melted by the heat of the impact. (Photos: Beda Hofmann.)

the level of the sand sheet. This conclusion was based on the paucity of calcium in the impact melt material. Had deeper rocks been melted, then relatively high levels of calcium should have been present in the impact material but weren't. Contrary to the results of Shoemaker and Wynn, the Swiss team concluded that the Wabar meteorite had been travelling from south to north. Previously, samples taken from the Wabar craters had been dated as 290 years old (with reference to a 2004 publication date), with an error of plus or minus 38 years [18]. This indicated that the Wabar meteorite fell sometime between 1676 and 1752 AD. These young ages, when combined with the revised flight direction of the fireball, led to a very significant conclusion. It now appears likely that the meteorite seen in the skies over Yemen on 1 September 1704 was actually the one that formed the Wabar craters [19].

So, what sort of meteorite is Wabar?

Well, not surprisingly, as it is largely made out of iron, Wabar is a type of iron meteorite. As we saw earlier, meteorites are often, and quite straightforwardly, divided into three broad types: (1) stones, (2) irons, and (3) stony-irons. The meteorites we met in earlier chapters were all stones, and in particular, a type known as chondrites (see Appendix 2 for further details). Iron meteorites are the ones most people think of as being typical space rocks. They just look like big lumps of iron metal, which is what they are! The idea that irons are "normal" meteorites certainly comes from the fact that they are the star exhibits in any museum meteorite display (Figure 5.7). They are often exhibited as big spectacular chunks and when cut and polished have beautiful, shiny surfaces. They look gorgeous! While their prominence in museums might suggest that they are a common sort of meteorite, this is not the case. Out of the 73,483 (February 2024) valid meteorite specimens listed on the Meteoritical Bulletin Database, only 1,365 specimens are iron meteorites. That's just under 2%. If you take a look at the Meteoritical Bulletin Database entry for the Wabar iron meteorite [20], you will see that it is listed as a type IIIAB. That needs a bit of further explanation.

It turns out that the best way to sort iron meteorites into their respective groups is to use the concentrations of a number of key elements. The best ones are nickel (Ni), iridium (Ir), gallium (Ga), and germanium (Ge). By analysing these key elements, it has been established that the vast majority of iron meteorites can be slotted into between twelve and fourteen principal groups, depending a little bit on how you lump and/or split them [21,22]. There are other irons that are not so easily pigeon-holed, and these are known as ungrouped irons. Of the grouped irons, the one that Wabar is a member of, the IIIAB group, is by far the most populated, with 349 recognised specimens (February 2024). Because some different specimens are likely to be chunks from the same meteorite, there are probably about 220 distinct meteorites in the IIIAB group.

But what does this all mean?

Looking at the chemical trends within individual iron meteorite groups suggests that most formed from molten metal. In an analogy to volcanoes on Earth which are fed by "magma", the molten irons are often called magmatic irons. Three of the iron groups don't seem to have formed from a molten metal liquid in any simple way and so they are known as the non-magmatic irons. But where would you get the conditions to form the magmatic groups? Like the Earth, which is believed to have a core consisting in part of molten metal [23], the magmatic iron meteorites formed at the heart of hot, ancient asteroids. Each magmatic group probably comes from a single asteroid. The evidence from the magmatic irons suggests we have samples of between 9 and 12 such mini-planets and if the ungrouped ones are also included, that number goes up to about 60 source bodies [24]. These melted asteroids would have started life very soon after the formation of the Solar System and would have been

FIGURE 5.7 The Nantan iron meteorite on display at the Oxford University Museum of Natural History. Always the star of a museum display, iron meteorites are actually quite rare specimens. (Photo: the author.)

composed, at least initially, of material similar to the stony meteorites we met earlier in the book. These early-formed asteroids would have contained radioactive elements that are no longer around today. Due to the heat given out as these radioactive elements decayed, the asteroids began to heat up. And in a similar way to an industrial blast furnace, as the rocks became molten, the dense metal they contained would have sunk to their core regions leaving the outer layers largely metal-free. In a very short time, these asteroids would have consisted of an outer rocky layer with a dense molten metal core. The fate of these old asteroids was varied. Some hung around for a long time, cooled and crystallised, before being "battered to bits" [24] and others were "battered to bits" while still largely molten. But all of them were eventually smashed up by numerous impacts, with some of these pieces arriving on Earth as fragments, just as the Wabar meteorite did. To sum this all up, iron meteorites like Wabar are smashed-up material from the core region of long disrupted asteroids.

Finally, we come to a rather perplexing question. If the Wabar craters only formed in 1704, how did they get mixed up with the legend of an ancient lost city? The Bedouins that accompanied Philby found it hard to accept that what they saw at Wabar was just a natural phenomenon and that the craters were not structures created by man. But building a legend around something that seems strange and disturbing is nothing new. Rainbows were linked to Noah in the Old Testament because, lacking a basic understanding of the physics of light, there was no way to make sense of these beautiful natural phenomena. The Wabar legend was perhaps borrowed as a ready-made way of explaining the impact craters and their associated iron fragments.

And finally, should we be looking elsewhere for the fabled city of Wabar? Philby, of course, had his own views on this. In a paper that he read to the evening meeting of the Royal Geographical Society on 23 May 1932, he said the following [25]:

> The discovery of such craters in a position so unexpected was sufficient compensation for the failure to find the ruins of a great city, but we had tracked the rumour of its ruins to their source, and I at any rate have done with the search for Wabar. That quest I bequeath to younger folk with the warning that now there is no more reason to suppose that any city ever existed in the vast expanse of the Rub' al Khali. The piece of iron to which I have referred, has provided the necessary clue to the character of the Wabar craters; and Dr. Spencer of the British Museum has been able to establish beyond any possibility of doubt that it is part of a large mass of meteoritic iron whose impact upon the earth gave rise to the craters.

Philby, of course, didn't have the last word on this because in 1992 "the younger folk" to whom he referred, in the form of Sir Ranulph Fiennes, believed they had located Wabar (Ubar) in the Dhofar region of Oman. And if that possibility turns out to be a deception, will the quest end there? Probably not! The search for Wabar will certainly continue and one day, who knows, it may be located.

When I worked in Saudi, I knew nothing of Wabar and Philby and I didn't know much about meteorites either. I discovered Wabar thanks to the Swiss team [17] who sent me samples of the black desert glass that is strewn across the site (Figure 5.6). It felt strange to be analysing extraterrestrial-related samples collected in the Rub' al-Khali while sitting in a nice, air-conditioned lab in the UK. But the work also showed me another side to that mysterious desert. The lost city of Wabar, like all such legends, grabs the imagination. Did it ever exist? Are its ruins still out there? Or was Philby right and now there is no longer any reason to believe in it? The mystery remains unresolved, as it usually does!

In the preceding chapters, we have been looking at some of the more exotic aspects of meteorites. How they can cause havoc, disrupt lives, and generally be a bit of a nuisance. But now it's time to look at their more positive contributions to life on Earth. Without giving the game away, it's fair to say that we wouldn't be here without them. Let's take a closer look.

6 What Have Meteorites Ever Done for Us?

You might agree that meteorites are interesting. Lots of things are. But vital to our very existence! Surely that's stretching it a bit? But no, without these periodic cosmic visitors, life on Earth would not be possible. Here, we look at the evidence for the vital role played by meteorites in making our Earth a habitable planet.

It's a beautiful day. The Sun is shining. You are having a relaxing coffee and a nice piece of cake on the patio. Lemon drizzle perhaps? All is calmness and tranquillity. You contemplate the possibility of a second cup of coffee, and why not a second slice of cake? It's too tempting to resist. So, you don't. But everything in this scene of convivial bliss is not quite what it seems. Because, in reality, you are pinned by gravity to the surface of a giant ball of rock and molten metal that orbits, at an average speed of 67,000 mph, around a natural nuclear reactor we call the Sun. By the way, the molten metal at the heart of our planet produces a strong magnetic field that helps to repel those high energy particles coming from the Sun and beyond. If that magnetic field wasn't there, you would certainly be fried as you sat there drinking your latte. There is always a lot more going on, even in the most tranquil of settings.

When we contemplate our middle-aged Solar System, it all looks so orderly. The Sun at the centre with its eight orbiting planets, moving as if they are part of a clockwork model. How nice, how regular, how predictable. But it wasn't always like this. For starters, where did it all come from? How do you build a planet? How do you build a Solar System? Let's look at how it all got going. It turns out that asteroids and meteorites are key players in this story. And it all began with dust [1].

Out there in the depths of space are areas enriched in gas, ice, and dust. They are sometimes called molecular clouds or dark nebulae (Figure 6.1) and are the birthplace of the stars and planets [3]. But to go from grains of dust to a star with orbiting planets is quite a journey and involves many complex steps [4,5]. Most schemes agree (a bit) on the main stages in this process, which starts with the gravitational collapse of a portion of the molecular cloud. The collapsing cloud forms a rotating disc of gas, dust, and ice. At its heart is a growing central object, which in the case of our Solar

DOI: 10.1201/9781003174868-6

FIGURE 6.1 NASA's James Webb Space Telescope reveals emerging stars and stellar nurseries in the Carina Nebula. The dark clouds are composed of gas and dust and contain newly formed stars [2]. (Image: NASA, ESA, CSA, and STScI.)

FIGURE 6.2 Artist's illustration of a protoplanetary disk [6]. (Image: ESO/L. CALÇADA.)

System, will eventually become the Sun (Figure 6.2). This disc is generally referred to as a protoplanetary disc, and such structures are commonly observed surrounding many young stars.

The Atacama Large Millimetre/submillimetre Array observatory, usually shorten to just ALMA, based in the northern Atacama Desert of Chile, has taken some stunning images of protoplanctary discs (Figure 6.3) [7].

FIGURE 6.3 Image of a protoplanetary disc surrounding the young star HL Tauri taken by the ALMA observatory. The dark zones within the system may mark areas in which planets are forming [8]. (Image: ALMA ESO/NAOJ/NRAO.)

During the very early stages of their development, protoplanetary discs do not contain planets or asteroids, just dust and gas. The dust slowly accumulates at the mid-plane of the disc. Material also spirals onto the growing and evolving central star. Gradually, the dust grains in the disc start to stick together to form clumps. And the clumps clump together to form ever bigger clumps. Centimetre-sized objects, then tens of centimetre-sized objects, and finally metre-sized objects are formed [4,5].

But once you get to metre-sized objects, scientists who model these processes [4,5] have found that there is a barrier to further growth, and this they call, not very poetically (well they are scientists!), "the metre-size barrier problem". Calculations and models suggested that once these growing lumps of material got to this sort of size they either smashed into each other and broke up, or alternatively drifted rapidly into the Sun and were lost. But as there are quite a few objects in the Solar System that are larger than one metre, there had to be an explanation. And there was. A process known as "pebble accretion" appears to get round this problem [9]. Pebbles are small centimetre-sized clumps of material which can accumulate in certain regions of the disc and then grow rapidly to form larger bodies [9]. It appears that there is no longer a barrier to growing larger bodies from dust and gas. Which is nice because we need our protoplanetary disc to be able to grow objects the size of Jupiter.

It is important to note that not all areas of the disc were behaving in the same way, at the same time. This is not so surprising as there would have been a strong temperature gradient away from the Sun. Compared to the inner Solar System, the gas giants, Jupiter and Saturn, would have grown much more quickly, trapping gas from the disc before it dispersed [10,11].

As has been demonstrated in the case of planetary systems around other stars (these are known as exoplanetary systems), giant planets can migrate from their point of formation [12]. This means that planets like Jupiter and Saturn may have formed further out in the disc and then migrated inwards (Figure 6.4).

When we look at our Solar System today, it is clear that it consists of two very different parts. The inner Solar System, where we live, and which contains the so-called "terrestrial planets" Mercury, Venus, Earth, and Mars. And the outer Solar System, with the gas and ice giants. The two regions are separated by a zone of rocky material known as the asteroid belt (Figure 6.5). As an aside, the asteroid belt is usually shown in movies to consist of a dense zone of smashed up rubble. Actually, it's a pretty empty place, and spacecraft can generally fly through it without ever crashing into anything, which is very reassuring.

We tend to think of the two zones of the Solar System as being separate from each other, as the distances between them are so enormous. At its closest, Jupiter is 365 million miles from Mars, the outer planet of the inner Solar System. However, as we will see, the evolution of these two zones has been closely linked from the very earliest stages of their formation. But for the moment let's just look in a bit more detail at how the planets in the inner Solar System formed. Well, it is the bit we live in, so worth just taking a few moments to find out how it got to be the nice, calm and orderly place it is today.

FIGURE 6.4 Research on exoplanets has revealed that Jupiter-like planets, which form in the outer cold regions away from their stars, can then migrate inwards. These giant planets that orbit close to their stars are called "hot Jupiters". This artist's concept is based on research using NASA's Kepler Space Telescope and suggests that this inward death spiral ultimately stabilises before the planet is destroyed [13]. (Image: NASA/JPL-Caltech.)

FIGURE 6.5 Artist's concept of our Solar System*. The Solar System as it is today consists of two very different parts. An inner region containing the so-called terrestrial planets (from left to right: Mercury, Venus, Earth, and Mars) and an outer region containing the gas and ice giants (from left to right Jupiter, Saturn, Uranus, and Neptune). The two regions are separated by a zone of rocky debris called the asteroid belt [14]. *not to scale – but I am sure you realised that anyway. (Image: NASA.)

At the end of the first stage of planet building, the inner Solar System would have been teeming with the so-called "planetesimals". These bodies would probably have been no bigger than about 100 km diameter and many would have been much smaller. Planetesimals were really just asteroids. But the difference between them and their modern counterparts is that these ancient asteroids would not survive for long. To get an idea of how many of these mini-planets there may have been, we just have to consider how many of them would have needed to merge together to create our Earth. Yes, that's right, big planets grow by sweeping up smaller ones. The biggest body in the asteroid belt today is Ceres. It is a fascinating world and was studied in detail by the NASA Dawn spacecraft in 2015 (Figure 6.6).

Technically, Ceres is classed as a "dwarf planet" and so is probably somewhat larger than these ancient planetesimals. But let's use Ceres for the moment as a stand-in for that first generation of mini-worlds. Ceres has a diameter of 939 km, so not so small really, and a mass of 9.4×10^{20} kg, which is a huge number.

Very big numbers like this are not easy to deal with and, for very good reasons, are almost always written in the format used here, known as "scientific notation",

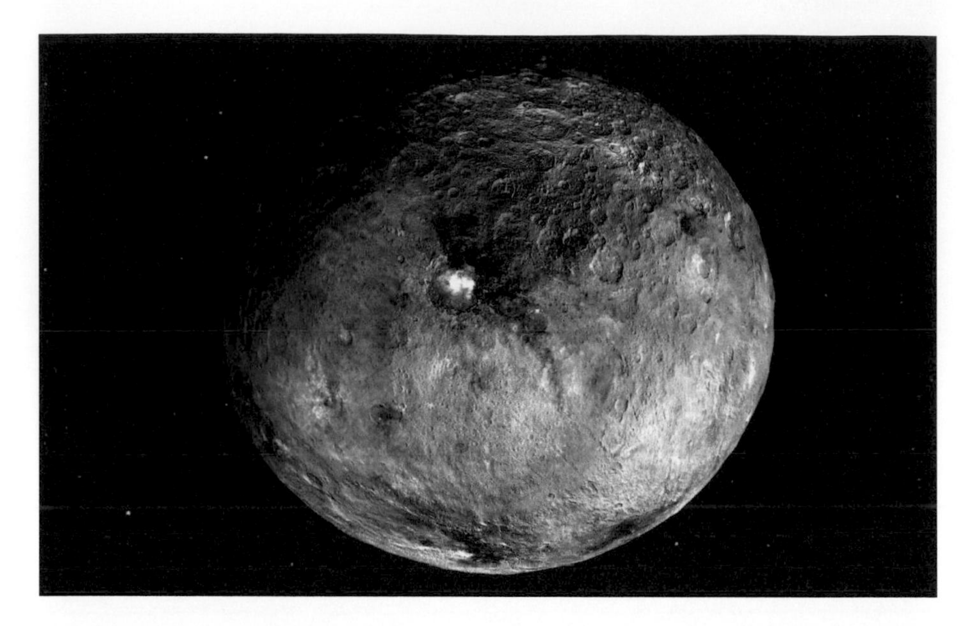

FIGURE 6.6 Dwarf planet Ceres shown in false colours to highlight differences in surface materials [15]. The bright material in the centre of the image is found in the large Occator impact crater and is thought to be salty deposits. Ceres likely contains liquid brines deep in its interior that are periodically released at the surface, probably as a response to an impact event [16]. (Image: NASA.)

9.4×10^{20}kg in words is nine hundred and forty quintillion. Yikes! What counts is the superscripted number at the end. If you want to appreciate the number of zeros involved, just bounce the decimal point to the right by the number of places given by the superscripted number. In this case, it's 20. For Ceres that means that 9.4×10^{20}kg becomes 940,000,000,000,000,000,000 kg, which is a mind bogglingly huge. So, in the end, scientific notation is the easiest way to deal with such crazy numbers.

The mass of Ceres might sound a lot, but compared to our Earth, Ceres is tiny. Our home planet has a diameter of 12,742 km and weighs in at a not insignificant 5.97×10^{24}kg. To put this into perspective, it would take 6,351 Ceres-sized asteroids to build the Earth. You can see that the inner Solar System must have been a pretty crowded place, teeming with tiny worlds. And we also have to build Mercury, Venus, and Mars, in addition to Earth. These thousands and thousands of mini-worlds would no doubt have orbited the Sun in a regular fashion, in much the same way as the larger planets do today. But due to the sheer number present, collisions would have been inevitable. Gradually, the number of planetesimals would be reduced. There would have been a few winners and many losers. The winners gradually increase in size at the expense of their nearest neighbours. Planetary scientists who study these sorts of things, call the area around a growing planet its "feeding zone" – nice! [4,5] (Figure 6.7).

FIGURE 6.7 Ceres, the largest body in the asteroid belt (bottom left), compared to the Moon (top left) and Earth (right) [17]. (Images: NASA.)

It would be survival of the biggest as the smaller bodies became incorporated into the larger ones. These bigger objects were not yet planet-sized, so they are generally referred to as "planetary embryos". The length of time involved in forming these embryos would have varied according to their location in the disc, but it was likely to have been between 100,000 and 1 million years [5]. And how large were these embryos? Well, they are usually said to have been Moon-to-Mars-sized [5].

You might think it strange that Mars could be referred to as a "planetary embryo" and not a full-scale planet. Of course, Mars is technically a planet, it's just a lot smaller than either Earth or Venus. And that's a bit of a problem. Mars has a diameter of 6,779 km, so about half that of Earth, and has a mass of 6.39×10^{23} kg, which does sound a lot. But here's the thing. You would need to combine just over nine planets with the mass of Mars to make our Earth. Perhaps that doesn't sound particularly strange. Why should planets all be about the same size anyway? Well of course, as we look around the Solar System today, with the notable exception of Venus and Earth, the planets are all very different in size. But these size differences give clues as to how each planet might have formed. When scientists have tried to model the evolution of the inner Solar System, they end up with all sorts of possible configurations. But the small size of Mars has remained a persistent puzzle [18].

Computer simulations of how the inner Solar System might have formed generally indicate that where Mars is found today there should be a bigger, at least Earth-sized, planet [18]. So, where is it? This has turned out to be such a profound question that it has changed the way we understand the early evolution of our Solar System. The

FIGURE 6.8 Planet Mars – its comparatively small size compared to Earth and Venus has been a long-standing mystery in planetary science. One possibility is that its "feeding zone" was depleted by the migration of Jupiter [18,19]. (Image: NASA.)

explanation may involve migration of the giant planets, the sort of thing that has been proposed for exoplanet systems (Figure 6.4). One very influential model that seeks to explain the small size of Mars is known as "The Grand Tack" [18], an allusion to sailing boats moving forward against a headwind. In the Grand Tack model, it is proposed that the giant planets, Jupiter and Saturn, first headed inwards towards the Sun reaching the point where Mars is now located and then moved outwards again to their current locations. The effect of this would have been to clean out the inner Solar System of planetesimals up to and including where Mars would eventually form. As a consequence, Mars would be smaller due to an impoverished feeding zone [18] (Figure 6.8).

The final stage in the development of the inner Solar System involved collisions between planetary embryos resulting in the formation of the large terrestrial planets, Venus and Earth. At this end stage, such collisions would have been highly energetic as there was no longer any gas around to dampen things down. These collisions were between bodies that were at least as large as Mars and so would have been spectacularly energetic events. And probably the most spectacular of all was the one that produced the Moon. Long known as the "Giant Impact", it was an event of cataclysmic proportions. The "classic" version of the Giant Impact envisages the collision of an object the size of Mars with the early, almost fully formed Earth [20]. The energy involved was so enormous that it almost totally melted our planet and produced a molten debris disc encircling the Earth, from which the early Moon coalesced in a matter of weeks to months. It would have been quite a sight. More recent versions of this model even envisage the Moon forming in a matter of hours [21,22].

That just leaves Mercury, the planet closest to the Sun. Mercury is the smallest of the four planets in the inner Solar System, with a diameter of 4,879 km and a mass of 3.29×10^{23} kg. Mercury is 18 times less massive than Earth and half as massive as Mars. Compared to the other terrestrial planets, Mercury's core is disproportionately large. This has led to the suggestion that its outer layers were stripped off during a close encounter with another large planetary body [23]. However, a single impact event seems unlikely and so Mercury may have suffered multiple giant impact collisions [24].

As you can see from the preceding sections, planets don't just pop up fully formed, they grow by collisions between smaller bodies. We can view the arrival of meteorites on Earth as just the end stage of a process that has been ongoing since the formation of the Solar System. Even today, each arriving meteorite makes the Earth a little bit bigger, although the rate of growth is clearly a lot slower than in the very early days of its formation. So, the Earth was a winner in the competition to thrive and survive in the early Solar System. But as it gradually grew at the expense of the smaller bodies in its local area, it would have been an inhospitable, barren world. For much of its main growth phase, it formed from dry, reduced (metal-rich) material. One reason we know this is because, had it formed predominantly from wet, oxidised (metal-poor) materials, the Earth would not have had a large molten, iron-rich core. Like the asteroids we looked at in Chapter 5, we can think of the Earth as being similar to the products from a blast furnace. The material that it formed from was probably locally sourced and similar in composition to a group of meteorites known as the enstatite chondrites. These are very reduced and composed of up to 13% iron metal [25]. And as we have seen, the Earth was periodically melted due to high energy collisions. As happens in a blast furnace, the dense metal in the new material added to the Earth would also have sunk down deep into the planet to join previous batches of molten metal. The result was that the Earth's core would have grown as each new batch of material was added.

Melting of planet Earth took place to a particularly spectacular extent following the giant Moon-forming impact event (Figure 6.9). The outer layers of the Earth were totally molten, with the amount of melting so extensive that it is generally described as a "magma ocean". It was a vast sea of lava that covered the entire planet to a great depth. As we have seen, the metal-rich core of the Mars-sized body that struck the Earth would have sunk through this magma ocean and joined the metal core that had already been forming inside our planet. The separation of molten metal into the Earth's core had big consequences for the chemistry of the planet. During this process, precious elements such as platinum and gold would have been removed from

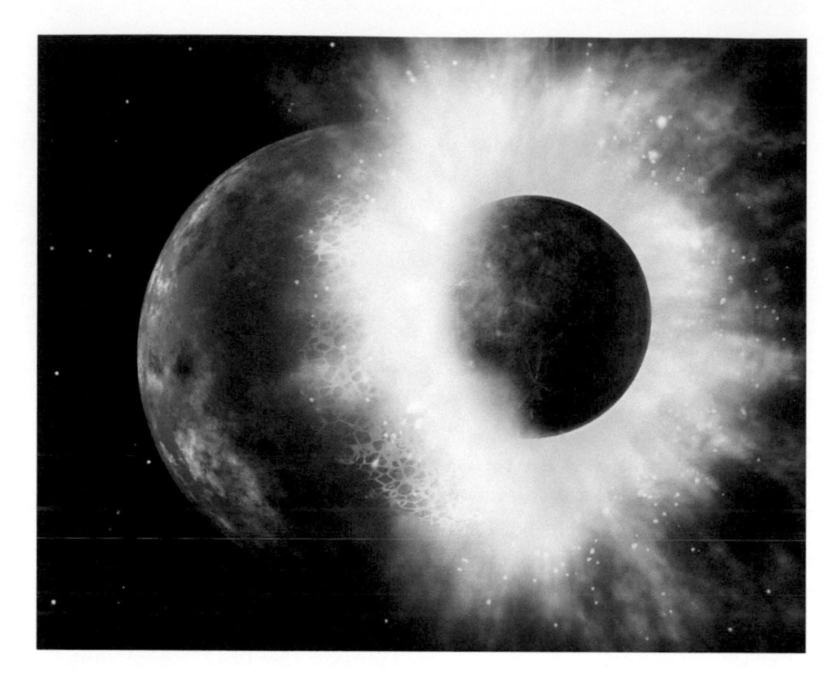

FIGURE 6.9 Artist's concept of a planetary collision. It is now generally accepted that the Moon formed as a result of an impact between the early Earth and another large body that may have been the size of Mars. The collision would have caused extensive melting of the outer part of the Earth producing a "magma ocean" [26]. (Image: NASA.)

the Earth's outer layers and have become incorporated into the core. Had things remained that way, all of the gold on Earth would have been locked as deep underground as it is possible to go. There would effectively be no gold left on Earth (apart from in the core). And yet we know from our everyday experience that gold, while not plentiful, is still available, at a price! (Figure 6.10). It turns out that we have meteorites to thank for all of the world's available gold reserves.

How do we know this? Elements such as gold, platinum, palladium, and iridium have a special affinity for metal. They prefer to hang out with iron-rich metal rather than silicon-rich rocky materials. Because of this, they are called highly siderophile elements or HSEs for short (siderophile literally meaning "iron-loving"). When these elements were analysed in samples from Earth's upper layers, there were two surprises. First of all, these elements were far more abundant than they should have been based on their affinity for metal, and secondly, all the different HSEs were present in similar abundances [28]. If they were just left-over elements after the metal had drained to form the Earth's core, their abundance patterns should have been highly irregular and that isn't the case. What was going on? The most straightforward and generally accepted way to explain these two features was if the HSE's had been added later, by meteorites [28]. This late addition of material would have taken place once the core was essentially closed off from the overly silicate-rich layers of the Earth. It is generally referred to as "late-accretion" and the material that was added could have taken place as a single event after the Giant Impact or more likely by a rain of meteorites that was more abundant than it is today.

FIGURE 6.10 Gold prospecting during the 1850 California gold rush. Gold has been one of the drivers of history. The opening up of the American West in the nineteenth century would have been a very different story in the absence of this precious yellow metal. And it is only available in findable quantities because of meteorites [27]. (Photo: Wikipedia.)

No matter how it happened, we have meteorites to thank for the world's available reserves of gold and other precious metals. Next time you contemplate the purchase of a beautiful set of gold earrings, it is worth remembering that they are essentially the distilled essence of meteorites – no wonder they cost so much! (Figure 6.11).

Delivery of precious metals to Earth's outer layers was an important achievement. But meteorites also provided us with a resource that is even more precious than gold. A substance essential for life itself. Water. We have seen how the Earth formed in the hot, dry inner Solar System. And yet Earth is a water world (Figure 6.12). Where did the water come from? It's a bit of a mystery.

The real stumbling block to working out where Earth's water came from is that we still don't know how much of it the Earth contains. Of course, there are the oceans that cover its surface, and these have an estimated mass of 1.4×10^{21}kg, which is a lot of course. This amount of water is said to be 1 "ocean mass". But rocks in the interior of the Earth also contain water, probably quite a lot. Some estimates put the amount of water on Earth at a staggering 18 ocean masses [29]. That's 2.5×10^{22}kg, enough to fill a lot of swimming pools! As we have seen, the inner Solar System was a dry place. That much water is unlikely to have come from a local source. One possibility favoured by many scientists is that the water was transported into the inner Solar System by water-bearing asteroids that formed in the outer Solar System [4,30].

An outstanding question is when did this water arrive on Earth? Not too early it seems otherwise forming Earth's metal core might have been problematic [31]. The

FIGURE 6.11 We have meteorites to thank for the world's reserves of gold. It is a metal that has changed the course of history. And is of course very beautiful too! (Photo: Marie-Christine Greenwood.)

FIGURE 6.12 As these chaps understand only too well, water is a vital component in making Earth a habitable and pleasant planet. Water was most likely transported to Earth by asteroids and meteorites during the later stages of Earth's formation. (Photo: Hélène Greenwood.)

idea is that Earth started to form from dry inner Solar System material but gradually some oxidised, water-rich asteroids started to creep into the mix. It wouldn't have needed that much to account for Earth's oceans. If we assume that Earth's exterior and interior contains the equivalent of 10 ocean masses (more than the minimum value of 1 ocean mass and less than the maximum estimate of 18), then only about 2% of the material that formed the Earth would need to have been provided by outer Solar System water-rich asteroids [32].

Next time you enjoy a nice cup of tea or put ice cubes into a long cool drink, it is worth remembering that the Earth is only a refreshingly habitable world, due to those water-rich meteorites that arrived too late for the main phase of Earth's formation but early enough to keep us all nicely hydrated – cheers!

FIGURE 6.13 Thanks to the asteroid that wiped out the dinosaurs, the Earth was left vacant to be inherited by the mammals, including us. T-Rex, Natural History Museum, Oxford. (Photo: the author.)

And finally, when discussing what meteorites have done for us, it is worth considering whether the human race would exist at all if it hadn't been for the impact event that wiped out the dinosaurs (Figure 6.13). Mammals were the great beneficiaries of this cosmic disaster and diversified to fill the ecological niches left vacant by the demise of the dinosaurs. Mammals also suffered as a consequence of the impact, but they eventually prospered from the catastrophe [33]. Had the asteroid strike not taken place mammals may still have outcompeted the dinosaurs, we will never know. But the rise of the mammals was given a cosmic boost by this event.

In this chapter, we have looked at some of the benefits of meteorites. We have seen that the Earth itself was built up from asteroids that were incorporated into it during its early growth phase. Meteorites from the outer Solar System are believed to have transported water into the inner Solar System, including Earth. We have meteorites to thank for those lovely gold earrings. And the dinosaur extinction event set mammals on a course to inherit the Earth. But there is one final benefit that meteorites bring us, and we will be looking at this aspect of the subject for the rest of this book. What we know about the origin and early evolution of our Solar System comes from the study of these lumps of space debris.

However, before we get into all that cool stuff, it's worth just asking the question: how did meteorites end up getting their crazy names? For example, what makes it OK to call a meteorite "Camel Donga"? Let's take a closer look at this very important question.

7 A Meteorite Called Camel Donga

Meteorites can have very strange names. How does that happen? Here, we look at one oddly named meteorite known as "Camel Donga" and find out that it has quite a tale to tell. What follows is one of the most successful detective stories in the history of meteorite studies.

Meteorites can sometimes have very bizarre names. Here are a few examples [1]: Bald Mountain, Dingle Dell, Git-Git, Mike, Ogi, Zag, Answer, Fuzzy Creek, Old Woman, Joe Wright Mountain, Licking, South African Railways, Bald Eagle, Nothing, Milly Milly, Aztec, and there are a lot more besides. So, how come meteorites can end up with such outlandish names? As we saw earlier, the naming of new meteorites is strictly regulated by the "NomCom" or to give it the full official title: The Meteoritical Society Committee on Meteorite Nomenclature. The NomCom publishes a clear set of guidelines, which make very interesting reading [2]. Section 3 of these guidelines: "New meteorite names" sets out how a meteorite should be named, and basically, it's all about the geographical features found close to where the meteorite was recovered. The specific rules involved are set out in Section 3.1 of the guidelines: "Geographic features":

> A new meteorite shall be named after a geographical locality near to the location of its initial recovery. Every effort should be made to avoid unnecessary duplication or ambiguity, and to select a permanent feature which appears on widely used maps and is sufficiently close to the recovery site to convey meaningful locality information....

Basically, the strange names some meteorites have been given reflects the fact that geographical names can sometimes be pretty weird. We are in big trouble if a meteorite ever lands close to the Welsh town of Llanfairpwllgwyngyllgogerychwyrndrobwl lllantysiliogogogoch [3].

You will often hear the curious and dubious fact that a meteorite takes its name from the nearest post office to where it was found, which is potentially a tough one for samples recovered in Antarctica or the Sahara. But the NomCom guidelines take this one head on: "*Two common misconceptions about meteorite names are widespread: that*

DOI: 10.1201/9781003174868-7

meteorites should be named for the nearest post office; and, that meteorites should be named for populated places such as towns. Neither of these is correct".

When it comes to strange meteorite names, my personal favourite is Camel Donga. It was found in 1984 on the Nullarbor Plain of Western Australia (Figure 7.1).

The official record for Camel Donga provides some basic information about the meteorite, noting that so far 25 kg of sample has been recovered [4]. The Camel Donga meteorite was discovered in January 1984 by Mrs J. C. Campbell during a trip with other family members travelling in two vehicles across the semi-arid limestone terrain of the Nullarbor Plain [5]. From the moving vehicle, they spotted the initial half-kilogram stone due to its brilliant black crust. A further eleven stones were recovered during a second visit to the area in July 1985 when Mrs Campbell was accompanied by two meteorite scientists, Brian Mason of the Smithsonian Institute and W.H. Cleverly of the Western Australia School of Mines [5].

The dark, glassy exterior found on almost all individual Camel Donga stones represents "fusion crust" (Figure 7.2). As we have seen, this is just a thin layer, generally no more than a millimetre or so in thickness which is produced as a result of frictional heating during the stones high velocity flight through the atmosphere. The interior of Camel Donga is actually pale in colour due to the presence of a high content of the light-coloured mineral plagioclase [5].

The strange name "donga" is a local Nullarbor term that refers to a shallow depression, generally covering an area of about 1 hectare [5]. The morphology of dongas means that they tend to concentrate local surface run-off and so can be more vegetated than surrounding areas. The presence of feral camel tracks was apparently the reason why Camel Donga acquired its distinctive name [5] (Figure 7.3).

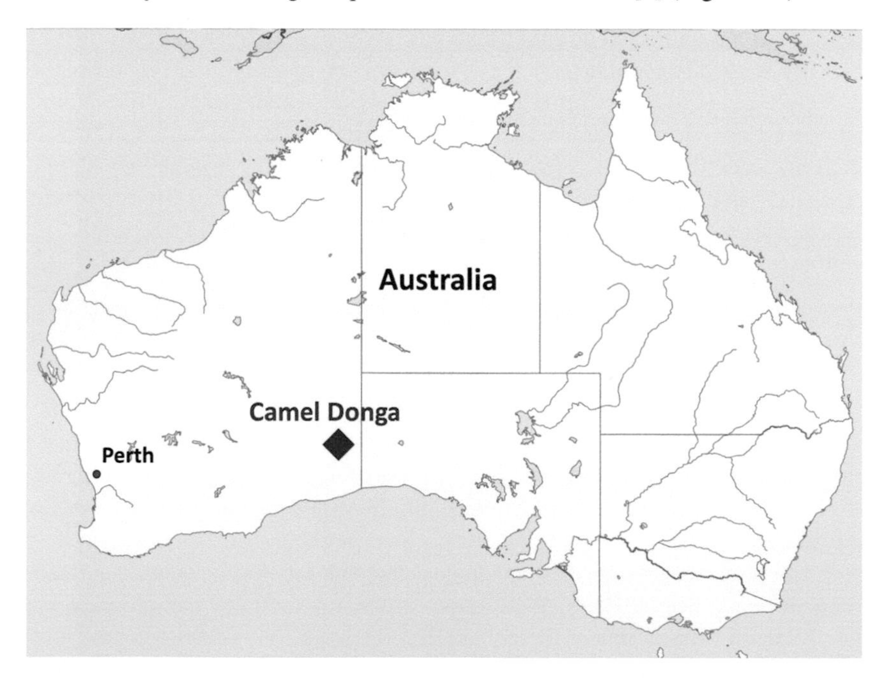

FIGURE 7.1 The find location of the Camel Donga meteorite in Western Australia. (Image: the author.)

FIGURE 7.2 A piece of the Camel Donga meteorite. The upper part of the sample is covered in a very thin layer of black, glassy fusion crust less than 1 mm thick. The light-coloured interior material can be seen on the bottom and right-hand edges of the specimen. (Photo: the author.)

FIGURE 7.3 Typical scenery in the Nullarbor region. (Photo: Andrew Morse.)

In fact, the Camel Donga meteorite is not the only one to carry this name. The Meteorite Bulletin Database currently lists 53 other meteorites which include "Camel Donga" as part of their official name. What's going on? There is a rational explanation, which is always nice! The location where the original Camel Donga was recovered is an example of a "Dense Collection Area". This is a fairly common situation in sparsely populated regions such as Antarctica, North Africa, parts of Australia, and North and South America. These are areas where large numbers of distinct meteorite samples have been collected but where there are few suitable geographical names. Special rules apply to such areas:

> If particularly numerous recoveries are made in one region, or are considered to be likely, as for instance in Antarctica and the Sahara, a generic prefix (conveying geographic information) and a suitable series of numeric suffixes should be applied. New meteorites found within the designated region will be named by combining the prefix with the next available suffix. [2].

Camel Donga was the first meteorite to be recovered from the Camel Donga area. There have been another 53 distinct samples found there since then. These are not just pieces from the Camel Donga meteorite but are compositionally distinct. The lastest, Camel Donga 054, was recovered in 2007 and is an ordinary chondrite [6]. The original Camel Donga should really be designated Camel Donga 001. But it was the first to be found and so is just known as Camel Donga, plain and simple.

Camel Donga has a lot more going for it than just a strange name and some shiny black fusion crust. The first clue to its origin comes from the classification details given in its official record [4]. Camel Donga is a eucrite, which means that it is predominantly composed of two minerals calcium-rich plagioclase and calcium-poor pyroxene. You might initially be disappointed to learn that eucrites are a relatively common type of meteorite. The Met Bull database currently lists the details of 1801 (February 2024) other officially recognised "eucritic" meteorites. But far from being disappointing, the fact that Earth receives so many meteorites with this composition is at first puzzling and at the same time a major clue to their origin. What follows is one of the greatest detective stories in the history of meteorite science. And you will be pleased to know it has a happy ending.

Eucrites, such as Camel Donga, are one of three related meteorite types that have become known by the acronym HED, the other two members being howardites and diogenites (**H**owardite, **E**ucrite, **D**iogenite = HED). Based on their composition and texture, eucrites are thought to represent volcanic rocks that formed either on the surface or close to the surface of an asteroid (Figure 7.4). Diogenites are thought to be related to eucrites based on their mineralogical similarities. However, the mineral grains in diogenites are generally larger in size than those in eucrites and so they most likely formed deeper in their parent asteroid and hence were able to cool more slowly (Figure 7.5). Howardites are smashed up rocks formed by impact processes and contain a mixture of eucrite and diogenite fragments. That's important because it suggests that all three meteorite types come from the same asteroid. But which asteroid might that be?

One way to match asteroids and meteorites is to look at the light that is reflected from their surfaces. The basics of this approach shouldn't come as too much of a surprise because we use the same technique every day of our lives. Analysing the light

FIGURE 7.4 Asuka 87272 is a 5.7 kg eucrite sample collected by the Japanese in Antarctica. It is partially covered in black, almost glassy fusion crust, which is a very characteristic feature of eucrites. Where the fusion crust has fallen away, you can see the very light-coloured interior. Like other members of the HED suite, eucrites are thought to be derived from asteroid 4 Vesta. (Photo: the author.)

that is reflected from objects is exactly what our eyes and brains do all the time. On a nice sunny day, you might be contemplating some beautiful scenery, and as a result, feeling a little bit more upbeat about life. But what you are actually doing is taking in the reflected light though your eyes and processing the image using your brain. It is more or less the same technique that is used to match asteroids and meteorites, it is just given the rather grand title of reflectance spectroscopy. When this approach was carried out on the HED meteorites, it was found that due to their unique composition their reflected light curves were very distinctive. In particular, they have a very characteristic absorption feature due to the presence of the mineral phase calcium-poor pyroxene [7]. For comparison purposes, it is also possible to analyse the light that is

FIGURE 7.5 Yamato 002875 is a 10.7 kg diogenite specimen recovered in 2000 by a Japanese team in the Yamato mountains of Antarctica. Its beautiful green colour reflects the fact that it is composed almost exclusively from large crystals of the mineral pyroxene. Diogenites are believed to be derived from deep in the interior of asteroid 4 Vesta. (Photo: the author.)

reflected from asteroids in space. And by comparing this reflected light to that from the meteorites, the two can be matched up. The aim is to try and locate the source asteroids of the meteorites landing on Earth. And that is exactly what has been done for the HED meteorites.

In 1970, Thomas McCord and colleagues reported the breakthrough discovery that the light spectrum from asteroid 4 Vesta, the second largest body in the asteroid belt, was a close match to that of the HED meteorites [8]. This spectral match has also been confirmed by other more recent studies [7,9] (Figure 7.6). Asteroid 4 Vesta, more commonly referred to as just Vesta, was discovered in 1807 by the German astronomer Heinrich Olbers. It was the fourth asteroid to be discovered, hence the designation 4 Vesta. The first "asteroid" to be discovered was Ceres in 1801 by Giuseppe Piazzi. Of course, Ceres is now classified as a dwarf planet.

But the fact that we get so many HED meteorites on Earth was something of a problem in linking them to Vesta. The asteroid belt is lumpy. Some parts of it contain relatively large numbers of asteroids while others are empty. This distribution is not an accident. The structure of the asteroid belt is very much controlled by its giant neighbour, Jupiter. Just like the planets, objects in the asteroid belt are in motion around the Sun. Depending on their radial position, some objects in the asteroid belt will regularly come closer to Jupiter than others. This situation is known as a resonance and tends to be unstable. The result is that certain zones in the asteroid belt contain very few objects because once something drifts into such a zone it tends to be ejected from the asteroid belt, potentially on an Earth crossing orbit. One of

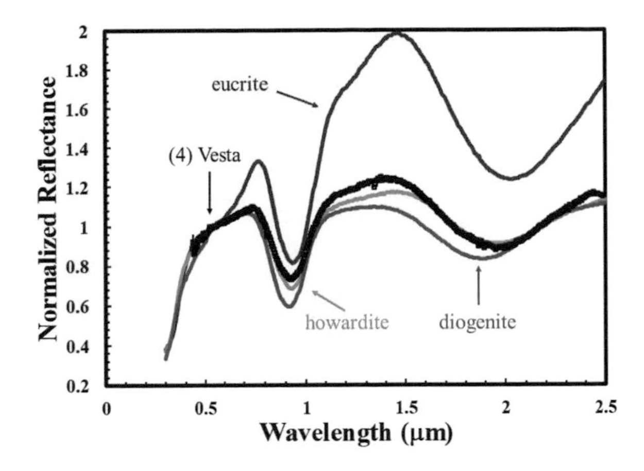

FIGURE 7.6 Reflectance spectrum of 4 Vesta compared to a eucrite, a howardite, and a diogenite. All the curves show a characteristic dip at about 0.9 μm and approximately 2 μm. These dips are known as absorption features and are a characteristic feature of the mineral low-Ca pyroxene. The similarity between the spectra for the HEDs and Vesta suggests that there is a genetic relationship between the two [10]. (Image: Tom Burbine.)

the most prominent gaps in the asteroid belt is the 3:1 resonance. This is where an asteroid would make three orbits of the Sun for each orbit that Jupiter makes [11]. The problem was that Vesta was not particularly close to such a resonance and so would, in theory, not be a good source of meteorites landing on Earth.

But then in 1993 came a breakthrough. Astronomers Rick Binzel and Shui Xu based at MIT identified a group of smaller Vesta-like asteroids, now known as "Vestoids" that bridge the gap between Vesta and the 3:1 meteorite producing resonance [12] (Figure 7.7). The Vestoids (as we shall see) likely formed by impact processes on Vesta. It was a mystery solved, and the connection between Vesta and the HED meteorites was now firmly established.

In 2001, NASA selected the Dawn mission as part of its Discovery programme with the twin objectives of undertaking detail orbital surveys of Ceres and Vesta, which are respectively the first and second largest bodies in the asteroid belt [13]. Testing the strength of the HED–Vesta connection was a stated science goal of the mission [13,14]. The plan was to visit Vesta first and then fly on to Ceres. Visiting two asteroids in one mission was ambitious and Dawn would make use of new technologies to achieve its goals. In particular, Dawn was powered by an innovative ion propulsion system [15]. Perhaps because it was such an ambitious mission, the spacecraft design and construction phase for Dawn was a pretty bumpy affair. And then on 2 March 2006, with the spacecraft largely complete, NASA decided to cancel the mission altogether [16].

A few weeks after the cancellation announcement, I attended the Lunar and Planetary Science Conference (LPSC) in Houston, Texas. At a routine NASA briefing held as part of the conference, the mood was very dark. Normally mild-mannered scientists were not happy, and a frank exchange of views took place with the NASA representatives. In particular, a German scientist affiliated to the Dawn mission didn't mince his words. He wasn't happy that the first he had heard of the cancellation

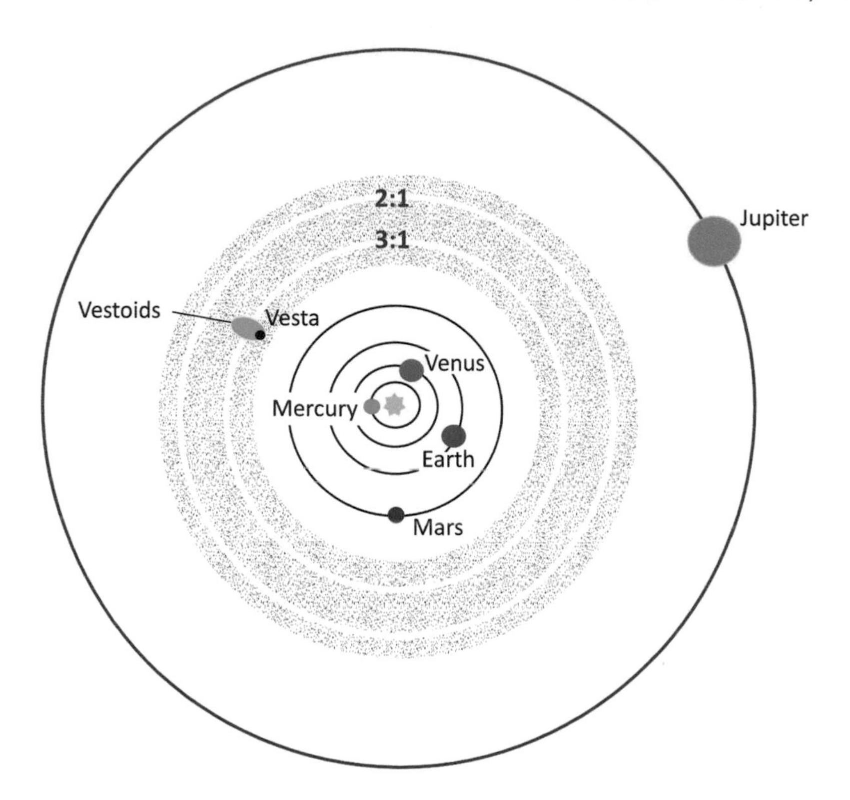

FIGURE 7.7 Inner Solar System and asteroid belt (grey stipple) showing the location of asteroid Vesta inwards of the 3:1 resonance and the zone of Vestoid asteroids which link the two. (Image: the author.)

was in the media. And there were many other scientists in the room who also voiced their displeasure [17]. Obviously, one angry meeting would not have been enough to change the cancellation decision and a lot must have gone on behind the scenes, but the result was that on 27 March 2006 the Dawn mission was reinstated. Phew!

The Dawn spacecraft was launched successfully in September 2007 (Figure 7.8). It went into orbit around Vesta on 16 July 2011 and undertook 14 months of detailed observations before departing for Ceres in the later part of 2012. It entered orbit around Ceres on 5 March 2015 and finally completed its mission on 1 November 2018 when its hydrazine fuel was essentially exhausted. The spacecraft is no longer operational and remains in a stable orbit around Ceres.

Dawn's observations and measurements essentially confirmed the HED – Vesta link beyond reasonable doubt [14,19]. The data returned by the spacecraft provided a very detailed picture of how Vesta had evolved. The images taken by Dawn show a world that came close to destruction due to the extent of the impacts that it experienced. Not only is it covered in an enormous number of small- to medium-sized craters, but the southern hemisphere of the asteroid is dominated by a very large double impact structure. The earlier impact structure is known as Veneneia and this is essentially overprinted by a second impact feature known as the Rheasilvia basin (Figure 7.9) [20].

FIGURE 7.8 The launch of the NASA Dawn mission to Vesta and Ceres on 27 September 2007 [18]. (Photo: NASA.)

A big surprise was the discovery of ridges and troughs close to Vesta's equator that probably formed during the impact events that created the Rheasilvia and Veneneia basins (Figure 7.10) [21]. The Rheasilvia impact is estimated to have taken place about 1,000 million years ago and the older Veneneia basin about 3,200–3,500 million years ago [21,22]. The Vestoid asteroid family likely represents debris ejected from Vesta during the event that formed the Rheassilvia basin. Dawn also revealed the presence of dark material on the surface of Vesta which most likely represents impact debris from carbonaceous chondrite-type meteorites [23]. Similar material has been found in the howardite meteorites. This feature is further evidence in favour of the link between Vesta and the HEDs. Estimates of the amount of eucrite compared to diogenite exposed on the surface of Vesta are also broadly consistent with their abundances amongst the HED meteorites that arrive on Earth [19].

Despite its prelaunch technical and financial issues, Dawn turned out to be a highly successful mission that provided critical information about asteroid Vesta and

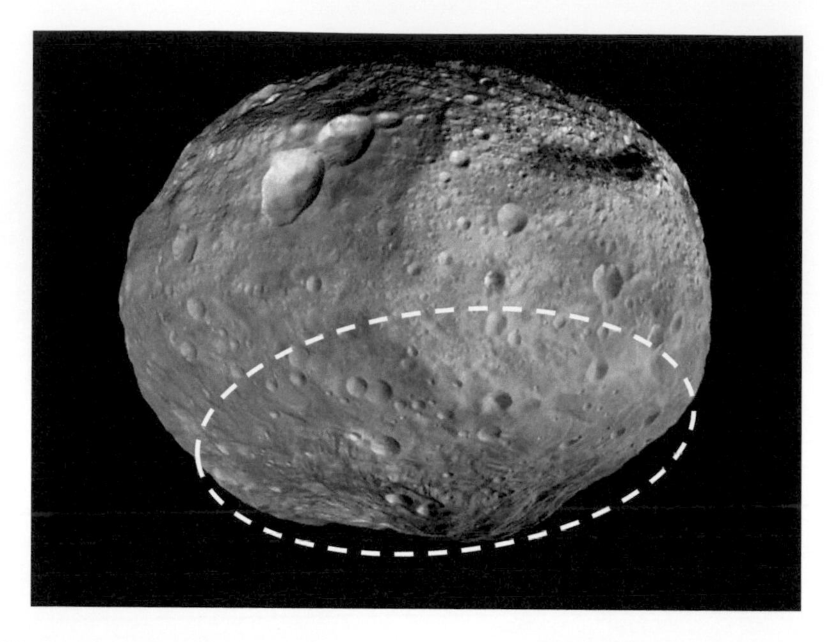

FIGURE 7.9 The surface of asteroid Vesta as imaged by the NASA Dawn spacecraft. The giant Rheasilvia basin (dashed outline) occupies much of its southern hemisphere and is thought to be the source of the Vestoid asteroids, which are responsible for the delivery of the HED meteorites to Earth. An earlier large impact basin underlies the Rheasilvia structure and is known as the Veneneia basin [18]. (Image: NASA/JPL-Caltech/UCLA/MPS/DLR/IDA.)

FIGURE 7.10 The Dawn mission revealed that Vesta has a series of ridges and troughs that run around its equatorial region. They probably formed during the impact events that created the Veneneia and Rheasilvia basins. Vesta survived these giant impacts, but only just [21]. (Image: NASA/JPL.)

its relationship with the HED meteorites. The results of Dawn's observations demonstrate conclusively that Camel Donga and all the other HEDs are escapees from Vesta. At its closest approach to Earth, Vesta is 170 million km away so Camel Donga had quite a journey. It was probably blasted off the surface of Vesta a billion years ago by the impact event that formed the Rheasilvia basin. It would then have been part of a Vestoid asteroid which drifted into the 3:1 resonance. Due to the strong tidal forces exerted by Jupiter, material that drifts into the 3:1 resonance doesn't stay there for long and so Camel Donga would have been rapidly launched on an Earth crossing orbit. Finally, the orbit of Camel Donga and that of the Earth would have crossed, and it would have arrived on the Nullarbor plain in a blaze of glory. Unfortunately, as far as we know, the fall of Camel Donga went unwitnessed. It is therefore classed as a "find", as opposed to a "fall" which is a witnessed recovery. But Camel Donga is a relatively fresh meteorite, and so, despite the fact that its arrival was never witnessed, it has probably not been on Earth for too long, a few centuries perhaps at most. And that as they say is that. Camel Donga is one of my favourite meteorites both because of its crazy name and its amazing story.

Unfortunately, not all meteorite mysteries have such a happy ending. If you appreciate a good old unsolved yarn like the Mary Celeste or the Roswell incident, then you will enjoy the tale of the Fer de Dieu meteorite – the giant space rock that vanished into thin air. Seriously, it did, (well sort of)!

8 The Tall Tale of the Soldier Who Found a Monster-Sized Meteorite and Then Lost It!

Meteorites can be very mysterious objects and none more so than the legendary "Fer de Dieu" or "Iron of God". But one mystery can lead to another. We begin our journey in the deserts of North Africa and end up back at the very dawn of our Solar System. In this tale, nothing is quite what it seems.

Everyone likes a good mystery story. Something that remains unexplained and perhaps also inexplicable. Even the most sceptical of us surely must admit that tales about the Mary Celeste, the Roswell incident, or good old Nessie hold a deep fascination. In fact, we often don't want to know the truth behind such stories. A little bit of us is happy with the thought that not everything in life has an easy or simple explanation. So, in the finest traditions of this genre, here comes the unsolved mystery of the lost "Fer de Dieu" ("Iron of God") meteorite.

It all started back in 1916 when a French colonial officer, Captain Gaston Ripert (Figure 8.1), claimed to have discovered a huge lump of extraterrestrial iron, measuring roughly 40×40×100m, in the desert of North Africa [1]. He said that the iron mass was located in sand dunes about 45km to the south-east of the town of Chinguetti in Mauritania [2] (Figure 8.2).

Captain Ripert claimed to have been taken to the spot in haste by a local guide and as a consequence had not been able to take any notes. He gave the following account of his first, and apparently only, visit to the huge meteorite mass [3]:

> ...In order to satisfy the urgent demands of my guide, I had taken with me neither compass nor any material that would permit me to make notes or any measurements whatsoever; also, I could remain only a very short time, because of the guide's haste to leave the spot; so, my observations were extremely cursory. I was able to state the meteorite

DOI: 10.1201/9781003174868-8

FIGURE 8.1 A photograph of Captain Gaston Ripert taken in the early 1940s. (Photo: Public Domain from Brigitte Zanda.)

> formed a sort of cliff … facing approximately southwest, while the northeastern side was completely buried in sand…. The upper surface was strongly polished by wind-blown sand. I found the small block up there, rounded on all its edges, that I delivered to M. Hubert. On one corner of the summit, facing west, I believe, erosion had exposed metallic needles sufficiently thick so that I could not break them or remove them. The metallic nature of the rock cannot be in doubt as shown by examination of the block I gave to M. Hubert. I tried to detach one of the needles by striking it very hard with the small block. The block shows traces of the shock at each point where it struck…. The surface appearance of the mass was in no way comparable to that of the blackish, polished surface of the rocks that are found on the 'reg' and on the sandstone plateau of the Adrar. On my return, I made some notes of my observations, while my recollections still were fresh; but I have since moved about so much that I do not know where any of my notes are…

As described in his account above, Ripert collected a 4 kg fragment lying close to the giant metal mass. A number of years later this fragment ended up in Paris and was identified as a relatively rare type of stony-iron meteorite, known as a mesosiderite (Figure 8.3). The fact that so much more of the meteorite remained in the desert was hailed as a great discovery. But a series of expeditions failed to trace the meteorite. In 1934, the search for the great "Fer de Dieu" meteorite was taken up by Professor Théodore Monod of the Muséum National d'Histoire Naturelle in Paris

FIGURE 8.2 A general view of the old town of Chinguetti in Mauritania [2]. (Image: François Colin/Wikipedia.)

FIGURE 8.3 Sample of the Chinguetti mesosiderite. It consists of a mixture of metal (grey, shiny blobs, and finer-grained grey material) and rocky silicate material (darker fragments and finer-grained material). (Photo: Brigitte Zanda.)

[1] (Figure 8.4). In the following 45 years or so, he undertook a series of expeditions, again without success. In 1989, he finally wrote:

> The existence of a giant meteorite in the Adrar of Mauritania, largely accepted since 1924, must now be abandoned. There was a mistake on the nature of the rock of a butte that is entirely sedimentary with no trace of metal [1].

FIGURE 8.4 Professor Théodore Monod of the Muséum National d'Histoire Naturelle in Paris who undertook a series of expeditions over a period of 45 years trying to find the Feu de Dieu meteorite, but without success. (Photo: Brigitte Zanda.)

Professor Monod concluded that the feature thought to be a giant iron meteorite by Ripert was in fact a sandstone and quartzite butte known as guelb Aouinet. However, Dr Brigitte Zanda, who accompanied Professor Monod to the area in 1991, remains unconvinced. In Ursula Marvin's 2007 account [1], Dr Zanda is quoted as saying that Ripert could only have mistaken guelb Aouinet as an iron meteorite if he had come across it "at night, at a distance, in a fog". At the 2011 Annual Meeting of the Meteoritical Society Meeting in London, I was able to talk briefly to Dr Zanda about the "Fer de Dieu". Dr Zanda had some insights into Captain Ripert's character based on his notebooks for other specimens he had sent back to Paris. Dr Zanda had the impression he was an honest and relatively knowledgeable observer of all things geological. She remained convinced that guelb Aouinet is not the locality that Ripert first encountered back in 1916.

But what about the meteorite that Ripert claimed was a sample of the larger mass. A study published in 2001 by Kees Welten and colleagues measured the concentrations of some key radioactive elements in the Chinguetti mesosiderite [4]. These unstable elements are called "cosmogenic radionuclides" and were produced in space when the meteorite was bombarded by a variety of energetic particles [5]. Once on Earth, the meteorite is then shielded by the atmosphere and these radioactive elements start to decay. Measuring the concentration of such naturally occurring radioactive elements can help to define how long a meteorite has been on Earth, and more importantly in the case of the" Fer de Dieu", how big the original object was before it entered the atmosphere. Dr Welten and his team concluded from their data that the Chinguetti sample came from an object that was no larger than 80 cm in radius prior to its arrival on Earth. This indicates that the sample that Ripert collected, and which eventually ended up in Paris, came from a very small meteorite. It seems it could never have been a piece from an object as large as the mythical "Fer de Dieu" mass as originally described by Ripert.

An additional problem is that a mass of meteoritic iron the size of the "Fer de Dieu" (40×40×100 m) should have produced a large crater and blown itself to pieces on impact with the Earth's surface. Thus, the iron meteorite which formed the 1.2-km-diameter Meteor Crater in Arizona is estimated to have been between 46 and 66 m in diameter prior to atmospheric entry (Chapter 9), a little bit smaller than the "Fer de Dieu" mass [6]. Based on its reported dimensions, Fer de Dieu would have had a mass of about 680,000 metric tons [7]. According to a paper by Philip Bland and Natalya Artemieva published in 2006, an impacting object of this size ought to form a crater with a diameter slightly greater than 1.5 km [8], or in other words, somewhat larger than Meteor Crater in Arizona. So, where is it?

The mystery of the Fer de Dieu meteorite remains unresolved, despite a significant amount of scientific research and considerable public interest [9]. It seems very unlikely that such a large object could have survived impact intact. In addition, the sample collected by Captain Ripert appears to have been part of a much smaller object. And yet what did he actually see that convinced him he had found a huge iron meteorite? The only way to completely solve the case would be to locate further fragments of the real Chinguetti meteorite. By finding such fragments in the desert, we might get a clue about the feature that Ripert mistook for a giant meteorite.

But there is an even deeper mystery related to the Feu de Dieu story and it is one which may have implications for how planets, including Earth, formed. As mentioned earlier, the piece of meteorite that Ripert sent to Paris for identification is officially classified as a mesosiderite (awful name, but there you go!) [10]. Mesosiderites are one of two meteorite groups that are collectively known as the stony-irons. As the name suggests, stony-irons are mixtures of metal and rocky material. They are also highly processed meteorites and could only have formed as a result of extensive melting in the asteroid from which they originated. So far, so good! But here is the mystery. In an asteroid that has been extensively melted, metal and rocky material should not co-exist as a scrambled mixture, as they do in mesosiderites. To appreciate this paradox, we need to delve a little into what happens when asteroids melt. They don't do this sort of thing much anymore, but back in the very early Solar System, melting was all the fashion. We will see why shortly.

The initial generation of asteroids that formed early in Solar System history were built from "primitive" dust (Chapter 6). This was a fairly complex mixture of materials, including a significant compliment of silicate-rich (rocky) grains and iron and nickel-rich metal particles. This dust also contained a very small but important component comprising short-lived radioactive elements [11]. These would have been produced in variety of different types of stars that predated the formation of the Solar System. So, when the cloud of dust and gas collapsed to form the protoplanetary disc (Chapter 6), it was already slightly radioactive. Of particular importance to our story is the short-lived isotope of the element aluminium known as aluminium-26 (^{26}Al) [11]. This decays to a stable isotope of magnesium, known as magnesium-26 (^{26}Mg). Aluminium-26 has a half-life of about 717,000 years [11]. The half-life is the time it takes for half of all the parent atoms present to decay to the daughter atoms. Although there will always be some parent atoms present, effectively after a few million years, most of the ^{26}Al would have decayed. Initially, at least the decay process gives out a lot of heat and would have been sufficient to almost completely melt the very earliest formed asteroids. This process can be modelled, and it has been found that an asteroid with an initial diameter of 130 km that formed 750,000 years after the start of the Solar System would have been completely molten just over 1 million years later [12]. Phonsie Hevey and Ian Sanders [12] summarise the situation nicely, pointing out that at this stage: "*the planetesimal is a globe of molten, convecting slurry inside a thin residual crust*".

Now here is the important part. As we saw in Chapter 6, metal has a much higher density than silicate-rich rocky material. So, once the asteroid melted, the metal would have sunk to the centre of the body forming a core, leaving the outer parts composed mainly of the silicate-rich material. But not all silicate minerals have the same density. This outer portion would also have separated into two layers. An inner coarser-grained layer consisting mainly of the minerals olivine and/or orthopyroxene [13] and an outermost, often finer-grained layer, containing a relatively high proportion of less dense minerals, in particular the mineral feldspar [14]. This results in a theoretical three-layer structure for these early asteroids, comprising a metal-rich core, an olivine-rich mantle, and a feldspar-rich crust (Figure 8.5). The crustal layer would have been fairly complex in structure and composition. Not only would it have comprised volcanic rocks similar to those we see today erupting from terrestrial

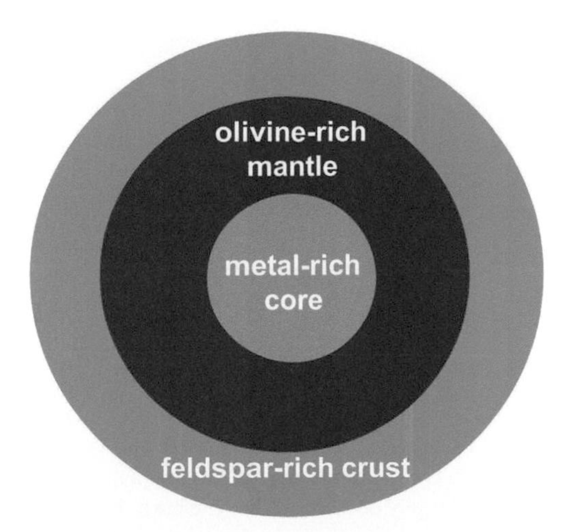

FIGURE 8.5 The internal structure of an asteroid that experienced near-total melting. (Image: the author.)

volcanoes, such as Mt. Etna or on Hawaii, but it would also have been modified by impact processes as the exterior of the asteroid was battered by meteorites.

And so finally, we come to why mesosiderites like Chinguetti are so enigmatic and puzzling. They consist of a complex mixture of metal and crustal silicate-rich material without any of the intervening mantle silicates. How do you do that if the asteroid started out with a three-layer structure? To make a mesosiderite, you would need to physically separate an asteroid into three bits: crust, mantle, and core, then mix the crust and core and throw away the mantle. At first sight that might seem like a tough, if not impossible job!

Sometimes to solve a problem, you have to look at it from a different angle. As we saw in Chapter 6, the planets essentially grew by devouring nearby asteroids. Computer simulation studies have shown that not all collisions between two asteroids would have resulted in perfect mergers [15] (Figure 8.6). Things could be messy. These simulations indicate that when a larger asteroid encountered a smaller one, the latter was often left in a highly deformed state. For example, its crust and mantle could be stripped off leaving a denuded molten metal core. Such encounters are known as "hit-and-run collisions" [15]. The two asteroids involved might eventually merge following a complex dance around each other, or they might just fly off in different directions. It was complicated (Figure 8.6).

But what has this got to do with the mesosiderites? As we have seen, they are complex mixtures of crustal and core materials without the intermediate mantle silicates. One possibility is that there was a hit-and-run encounter between two asteroids in which one of the two bodies lost its crust and mantle layers, leaving just a

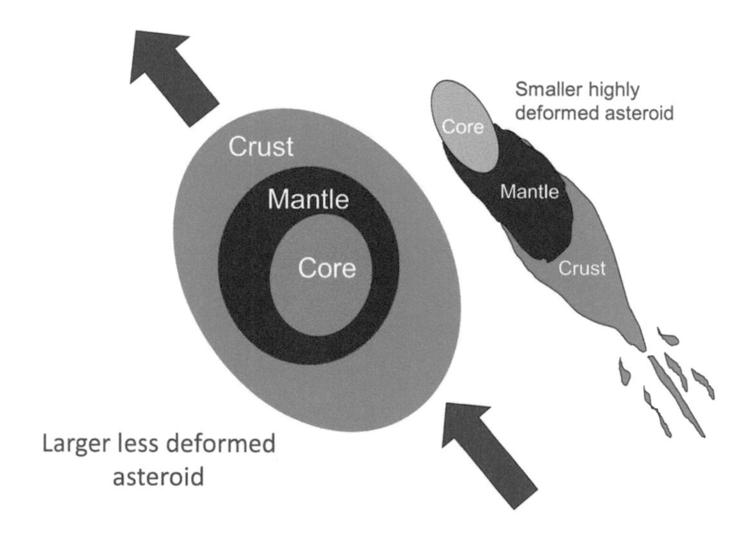

FIGURE 8.6 A schematic diagram showing one possible outcome following a hit-and-run collision between two asteroids [15]. The smaller asteroid is heavily deformed by the encounter, with the crust and mantle largely detached from the core. It is possible that the mesosiderites represent impact-related products formed when a denuded, partially molten iron core impacted the outer layers of a differentiated asteroid. It has been suggested that this asteroid might have been Vesta [16–18]. (Image: the author.)

FIGURE 8.7 Not really a hit-and-run collision! In this situation, the two "objects" have joined together to form a new single larger "entity". (Photo: Mark Errington.)

denuded metal core. Following this encounter, the now-denuded metal core impacted the crustal layers of a relatively intact asteroid and mixed with it to form the mesosiderites. A hit-and-run collision is not the only way to explain the mesosiderites and certainly other mechanisms have been proposed. But it has the advantage that it does seem to explain some of the puzzling features of these strange meteorites, which is always nice! [16].

But there is one further mystery about the mesosiderites which unfortunately remains unresolved. The rocky material found in mesosiderites bears a close resemblance to the eucrites and howardites of the HED meteorites. There are differences as well, which may have been caused by heating and melting of the colder silicates when mixed with hot, molten metal. But from the perspective of one critical line of evidence, HEDs and mesosiderites are pretty much identical. They both have the same oxygen isotope composition [17]. As we saw in chapter 4 (page 40), oxygen isotopes are an excellent way of tracing the relationships between different meteorite groups, and the fact that the HEDs and mesosiderites are so similar gives pause for thought. Could the mesosiderites be from Vesta? It was something the NASA Dawn team looked out for, but no supporting evidence has been found [19]. A number of recent studies have reached very different conclusions about the relationship between Vesta and the mesosiderites. Some studies have proposed that the mesosiderites were formed on Vesta [16–18], while another has rejected this possibility [20].

We started out this chapter with one mystery and have ended it with another! The mythical Fer de Dieu appears to be just that, a myth. But the possibility that the enigmatic mesosiderites come from Vesta remains unproven, for the moment. It is also clear that impact processes in the early Solar System were diverse. Some led to relatively straightforward mergers of colliding bodies (e.g., Figure 8.7), or in the case of hit-and-run collisions, a range of diverse outcomes are likely [15] (Figure 8.6).

But let's not forget that meteorite impacts can also have a very human side too. While they can blow a massive hole in the ground in an instant, they can also seriously impact a man's sanity and destroy his fortune. It is time to visit Meteor Crater, Arizona and find out how one man's obsession changed our understanding of the processes that take place when space rocks impact Earth.

9 Having a Smashing Time
The Story of Meteor Crater

This is the story of how one man thought he could make a fortune mining a meteorite, only to find it had vanished into thin air. We look at how obsession and financial disaster contributed to a better understanding of catastrophic meteorite impacts. Our story includes a clear case of attempted fraud, nuclear explosions, and massive asteroid bombardment. What not to like!

When an incoming meteorite is big enough, nothing can slow it down. Travelling at cosmic velocity [1], it will impact Earth with cataclysmic consequences. How do we know this? Now, that's a very interesting question. It took scientists a long time to understand the processes and structures that form in the aftermath of a full-scale meteorite impact event. Nothing illustrates this better than the controversies that have surrounded the best-preserved impact structure on Earth – Meteor Crater, Arizona [2].

Located 32 km west of the small town of Winslow [3], Arizona, Meteor Crater is the best known and almost certainly the best studied impact structure on Earth (Figure 9.1). It has a diameter of 1.2 km and is about 180 m deep [4]. The original size of the asteroid that produced the crater is estimated to have been between 46 and 66 m in diameter [1]. David Kring of the Lunar and Planetary Institute points out that a similar sized impact today would destroy an urban centre the size of Kansas City (population 508,000) [4,5]. Based on the extensive amount of research that has been carried out at Meteor Crater and the surrounding plains, no scientist today would seriously question its extraterrestrial origin. But that was not always the case, as we shall see. And to add a little spice to the story, it all started with some attempted skulduggery.

It is generally accepted that scientific interest in Meteor Crater really started with the work of Albert Edward Foote (1846–1895), who was a trained doctor and a chemistry professor at Iowa State Agricultural College (now Iowa State University) [6–9] (Figure 9.2). But, more importantly for our story, Foote also had a deep interest in mineralogy and had started a mineral dealing business in Philadelphia.

Foote's involvement with Meteor Crater was triggered by a sequence of events that started in March 1891, when a mineral prospector contacted the mining firm of N.B. Booth and Co. based in Albuquerque, New Mexico, claiming to have found a "vein of metallic iron near Canyon Diablo". A sample of the iron was sent with the letter and a request made for a chemical analysis ("assay") of the material. The analysis was

DOI: 10.1201/9781003174868-9

FIGURE 9.1 Meteor Crater, Arizona is the world's best-preserved impact crater. (Image: NASA.)

undertaken by a firm based in Colorado and the results indicated that it was mainly composed of iron (76.8%), containing some lead and silver, and traces of gold. The report on the sample suggested that it might be "furnace product", i.e., slag; in other words, a waste product from metal smelting. Understandably, the mining firm was not completely happy with these results and decided to seek further advice. They sent out two samples, one to the President of the Santa Fe Railroad and another to General Williamson, who was the land commissioner of the Atlantic and Pacific Railroad Company based in Chicago.

General Williamson contacted Albert Foote about the potential of a mine composed of "pure metallic iron" and provided the additional details from the prospector that:

> The vein had been traced for a distance of about two miles, that it was forty yards wide in places, finally disappearing into a mountain and that a car load could be taken from the surface and shipped with but little trouble [10].

It is clear that Albert Foote didn't take any of these details seriously. He knew a meteorite when he saw it:

> A glance at the peculiar pitted appearance of the surface and the remarkable crystalline structure of the fractured portion convinced me that the fragment was part of a meteoric mass, and that the stories of the immense quantity were such as usually accompany the discovery of so-called native iron mines, or even meteoric stones [10].

FIGURE 9.2 Albert Edward Foote identified the Canyon Diablo meteorite and studied Meteor Crater, but never firmly linked the two. (Photo: Iowa State University.)

Intrigued by the discovery, he decided to visit the locality himself, which he did in June of the same year. He was also not convinced about the chemical analysis and asked for further details. It turned out that the lead, silver, and gold were all likely to have been due to contamination introduced during the analysis process, which must have been a bit embarrassing for the company concerned.

Once at the field site, Foote established that the mineral prospector had been a bit liberal with the truth as: "*the quantity had, as usual, been greatly exaggerated*". But Foote, in print at least, puts a charitable spin on things: "*There were some remarkable mineralogical and geological features which, together with the character of the iron itself, would allow of a good deal of self-deception in a man who wanted to sell a mine*" [10]. So, our story kicks off with a failed attempt to pass a meteorite impact site off as a commercial iron mine.

Foote points out that most of the meteorite fragments were found close to "*the base of a nearly circular elevation which is known locally as Crater Mountain*". He goes on to provide an excellent description of Meteor Crater:

> The elevation, according to the survey, rises 432 feet (131.67 meters) above the plain. Its center is occupied by a cavity nearly three quarters of a mile (1.2 kilometers) in diameter, the sides of which are so steep that animals that have descended into it have been unable to escape and have left their bleached bones at the bottom [10].

Foote seems to imply that there must be some sort of relationship between the crater and the fragments but never overtly expresses this connection. In fact, he says:

> Careful search, however, failed to reveal any lava, obsidian or other volcanic products. I am therefore unable to explain the cause of this remarkable geological phenomenon. I also regret that a severe gallop across the plain had put my photographic apparatus out of order so that the plates I made were of no value.

Albert Foote summarised his studies at Meteor Crater in a paper he presented to the Association for the Advancement of Science on 20 August 1891 [10]. In addition to detailing the circumstances of his involvement with Meteor Crater, he also provided clear evidence that the iron fragments found around the crater were meteoritic in origin, containing 3% nickel, diamond and, in one of the accompanying figures, showing a clear example of Widmanstätten pattern [11]. The meteoritic samples described by Foote are officially known as Canyon Diablo, after a river gorge that is a few miles from the crater [12]. Canyon Diablo is an iron meteorite of the IAB group [13] (Figure 9.3).

In his comprehensive guide to the geology of Meteor Crater [9], David Kring notes that in the audience when Albert Foote presented his paper, was the chief geologist of the United States Geological Survey, Grove Karl Gilbert (1843–1918) (Figure 9.4). He was clearly fascinated by Foote's results and made a two-week trip to the crater in November of 1891 to take a further look at the problem. Grove decided to test two distinct ways in which the crater might have formed, either it was the result of a meteorite impact, or alternatively it was produced by a volcanic-related steam explosion [9]. He reasoned that if a very large meteorite was the culprit, then some remnants of the projectile should be buried beneath the crater floor. To test this idea, he conducted a detailed topographic survey of the crater with the aim of establishing its volume [9]. This measurement could be compared to the volume of ejected material forming the crater rim, and if the two were, more or less, the same, then there was unlikely to be a significant mass of meteoritic material hidden below the crater. Gilbert and his team also conducted a survey of the magnetic field around the crater, on the basis that a large mass of buried iron would be detected using this technique. The volume and magnetic tests failed to show any evidence in favour of a large mass of subterranean extraterrestrial metal at Meteor Crater, and so Gilbert concluded that the structure formed during a volcanic-related, steam explosion.

Gilbert, who was also President of the Geological Society of Washington, presented the results of his studies at Meteor Crater as a Presidential Address to the Scientific Societies of Washington on 11 December 1896 [14]. Sadly, Albert Foote had died a year earlier, on 10 October 1895, of chronic tuberculosis [7]. It is always possible that he might have disputed Gilbert's conclusions. We will never know. It seems that as a result of his eminent position, the steam explosion origin for the crater put forward by Gilbert was widely accepted at the time [9].

Gilbert provides some further interesting details about the prospector who originally forwarded the meteoric iron to the mining firm of N.B. Booth and Co. His name was Craft, and having some form of mining claim to the area, was actively

FIGURE 9.3 A fragment of the meteorite that created Meteor Crater, Arizona. It has been given the official name Canyon Diablo after a small nearby river valley, in which most fragments of the meteorite have been collected. No samples have ever been recovered from within the crater itself but were found scattered across the nearby plains. Canyon Diablo is an iron meteorite. Graham Ensor meteorite collection. (Photo: the author.)

negotiating a sale to these claims, but Gilbert notes: "*The negotiation was not concluded, because Mr. Craft, having borrowed money on the strength of his great expectations, mysteriously disappeared*".

Gilbert's conclusion that Meteor Crater was not formed by an impact process was all the more surprising because only a few years earlier he had presented a paper discussing the origin of craters on the Moon and concluded that they formed by meteorite impact [15]: "*The impact theory as thus developed appears competent to explain the origin of all the typical features of the lunar craters*". However, in one very important respect, Gilbert was correct – there was no large

FIGURE 9.4 Grove Karl Gilbert was in the audience when Albert Foote presented his paper on Meteor Crater. He was fascinated by what he heard and undertook his own studies into how it formed. Unfortunately, for the right reasons, he got the wrong answer and decided it had formed due to a volcanic steam explosion. It was a serious professional mistake, which Daniel Barringer would later highlight. G. K. Gilbert was a well-liked figure and the criticism of him by Barringer was not well received by fellow scientists at the time. (Photo: USGS.)

mass of meteoritic iron lying beneath Meteor Crater. It was a conclusion that the next important character in this story vehemently disagreed with and he would pay dearly for his mistake.

Daniel Moreau Barringer (1860–1929) was initially a lawyer, who then became involved in mining activities in which he made a considerable fortune [16] (Figure 9.5). A casual conversation in 1902 on the veranda of the San Xavier Hotel in Tucson with Samuel J. Holsinger, then of the US Forestry Service, made Barringer aware of the crater and the meteoritic iron that it was associated with [9]. Apparently, Barringer was shocked that he had never heard of it before and Holsinger recalled that he "dropped his cigar" in astonishment [17]. Barringer didn't do anything immediately but sometime later told the story to his friend Benjamin Chew Tilghman, a mathematician and scientist. Tilghman was also captivated by what he heard and they both decided to make further enquiries. In fact, they got Holsinger to undertake some research, and based on the results of this study, decided to acquire the mining patents to the crater. Barringer was a hunting companion of President Theodore Roosevelt, who personally signed the relevant paperwork! [16]. Once Barringer had obtained the property, he formed the Standard Iron Company

FIGURE 9.5 Daniel Moreau Barringer, who championed the extraterrestrial impact origin of Meteor Crater. His drilling operations at the crater were spectacularly unsuccessful and he lost a fortune. However, his radical views on the formation of the crater were posthumously vindicated by detailed scientific study. (Photo: Barringer Crater Company.)

and immediately began survey work and drilling operations [9]. In 1905, Barringer published the first results of this work in the Proceedings of the Academy of Natural Sciences in Philadelphia [18]. In the publication, Barringer provides multiple lines of evidence indicating that the crater formed by a meteorite impact and not a volcanic steam explosion as proposed by Gilbert. Barringer was far from polite about the work of Gilbert and states:

> Arguments which between us we have worked out, in support of the theory that this gigantic hole is an impact crater, will be set forth in the two following papers. It must be remembered that while a great deal of the evidence collected by us is positively in favor of the theory, much of it is negatively so; that is to say it disproves the theory that this great hole is the crater of an ancient volcano, or was produced by an explosion of steam, which latter theory seems to have been adopted by Mr G. K. Gilbert on what seems to be very insufficient evidence. Perhaps it would be more accurate just to say that he has adopted this theory because of an inadequate examination of the phenomena at Coon Mountain, or, as it is frequently called, Coon Butte; for had he examined the surface carefully, it does not seem possible to me that any experienced geologist could have arrived at such a conclusion.

Gilbert was a well-liked and respected geologist and this criticism of him didn't help to convince other scientists of the case for an impact origin for Meteor Crater. However, Barringer probably wasn't particularly worried by this, as his main goal was to locate the large mass of meteoritic iron that he was now convinced lay below the floor of the crater.

Drilling and survey work at the crater continued until 1928 and yet no trace of the giant buried meteorite was found. Huge sums of money had been expended to no avail. Not surprisingly, the directors of the Standard Iron Company got increasingly worried about the situation. An additional sum of $200,000 was raised for further work. By this stage, Barringer's estimate of the mass of the buried meteorite was in the region of a jaw-dropping 10 million tons, which would, if realistic, have resulted in a staggering quarter of a billion dollars profit [17]. But things didn't go well, and when the new mining operations hit a huge quantity of ground water, the patience of Barringer's fellow directors finally ran out. They commissioned an independent survey by the astronomer F.R. Moulton [17]. His analysis was devastating, Moulton's work indicated that due to the colossal amount of energy involved, the crater-forming projectile would have been completely vapourised. He concluded that there was little chance that any significant amount of meteoritic material remained buried under the crater floor. Within a week of receiving Moulton's second, more comprehensive report in November 1929, Barringer died of a massive heart attack and his dream of mining Meteor Crater died with him [17].

But in the years since his death, Barringer's initially unpopular interpretation of Meteor Crater as an extraterrestrial impact structure went from being a minority view to gaining widespread scientific acceptance. In fact, impact cratering is now seen as a major process that has significantly modified the surfaces of planets and asteroids throughout the Solar System [9,19]. But general acceptance of an extraterrestrial origin for the crater had been a slow process.

One important and enigmatic figure who carried out collecting and survey work at Meteor Crater was Harvey Nininger. Nininger, a self-taught meteorite collector and dealer, was one of the founding members of the Meteoritical Society [20]. In 1946, he was looking for somewhere to house his large meteorite collection and leased a property on Highway 66, about 5 miles from Meteor Crater, which became the American Meteorite Museum [20,21]. He met with members of the Barringer family and was given permission to search for and collect meteorites at the crater. He undertook a survey of about 23 acres of land close to the crater using a magnetic rake attached to his Studebaker car. By this method, he isolated 12,000 small meteorites [20]. He became particularly interested in the small soil particles that were attracted to the magnetic rake and which contained detectable nickel, a good indicator that they were of extraterrestrial origin. Nininger concluded that these particles were condensation products that formed from the meteoritic iron that had vapourised on impact, as first suggested by F.R. Moulton.

But Nininger's activities at Meteor Crater were terminated abruptly in 1948. Without consulting the Barringer family in advance, Nininger successfully petitioned the American Astronomical Society to pass a resolution calling for Meteor Crater to be taken into public ownership [20]. In support of his petition, he suggested that the Barringer family would be open to a sale provided they were offered a fair price. But this wasn't true. The Barringer family had always made it clear that Meteor Crater was not for sale. They were incensed by the petition and as a consequence, Nininger's permit to collect samples and undertake exploration at Meteor Crater was abruptly terminated. Nininger continued to operate his museum on Highway 66 until 1953, when he moved it to Sedona, Arizona [20].

Conclusive proof that Meteor Crater was formed by a high energy impact process finally came in 1960 when Edward Chao, Eugene Shoemaker (Figure 9.6), and Beth Madsen of the United States Geological Survey identified a mineral known as coesite in sandstones at the crater [22]. Coesite is a type of quartz that only forms at extremely high pressures [23]. Eugene Shoemaker (1928–1997) studied Meteor Crater for his PhD work at Princeton University. He noted that Meteor Crater displayed many similarities to the craters formed during atomic bomb tests carried out at the Nevada desert test site in the 1950s [24] (Figure 9.7).

FIGURE 9.6 Edward Chao (right) and Eugene Shoemaker (left) pose with some samples from Meteor Crater in which they identified the high-pressure mineral coesite. This work was undertaken with their United States Geological Survey colleague Beth Madsen [22,23]. (Photo: USGS.)

FIGURE 9.7 The Sedan Crater (foreground) was formed on 6 July 1962 by a 104 kiloton nuclear detonation. As highlighted by the studies of Eugene Shoemaker [24], there are similarities between the structures present at Meteor Crater and those formed during nuclear explosions. (Photo courtesy of National Nuclear Security Administration/Nevada Field Office.)

But why should nuclear and meteorite craters show close similarities? After all, the nuclear craters formed by detonation of bombs that had been drilled into the ground and meteorite craters are produced by objects travelling at very high speed. The similarity between the two types of craters is a reflection of the fact that both formed as a consequence of explosions. While the meteorite may have been travelling at high speed prior to impact, it was essentially stopped dead by the surface that it ploughed into. The huge amount of kinetic energy that it had prior to impact was then, more or less, instantly transferred from the projectile to the surrounding rocks.

The two nuclear craters highlighted by Shoemaker, Jangle Uncle and Teapot Ess, were actually quite small, with Teapot Ess being about 90 m wide and 36 m deep [25]. The amount of energy released in the Teapot Ess explosion was equivalent to about 1,200 tons of TNT. In comparison, the impact event that formed Meteor Crater was likely close to 10 million tons of TNT [26], more than 8,000 times more powerful than at Teapot Ess. A much larger explosion was carried out on 6 July 1962 at the Nevada Test Site, using a nuclear device buried 193 m underground. The resulting explosion had an energy equivalent to 104,000 tons of TNT, still nearly one hundred times less than at Meteor Crater, but closer than Teapot Ess. The test produced an

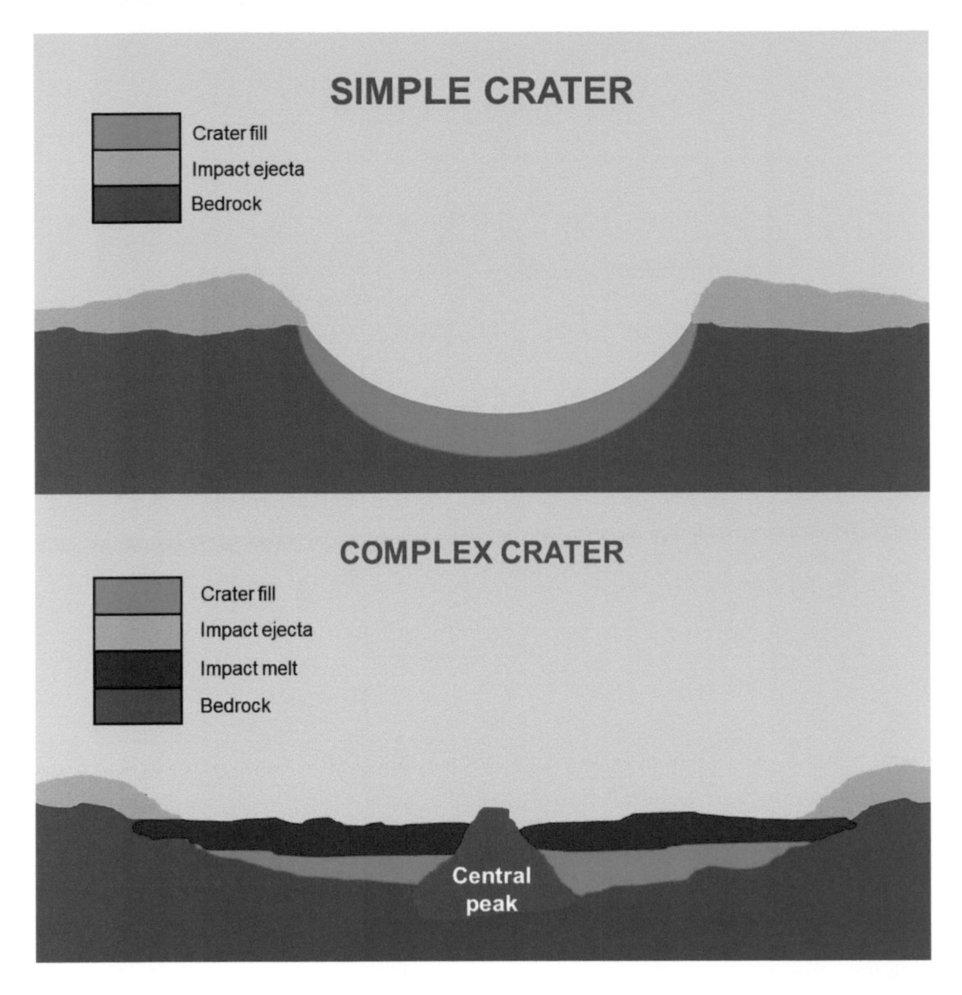

FIGURE 9.8 (top) Smaller craters on both the Earth and Moon, typically with diameters of less than about 15 km, have a bowl-shaped cross-section, and are known as simple craters. (bottom) In contrast, larger craters have a more complex structure with a central uplift peak, flat floors, and extensive development of impact melt sheets. (Image: the author, modified after [28,29].)

explosion crater known as the Sedan Crater, which is 390 m wide and 91 m deep, and bears a remarkable resemblance to Meteor Crater (Figure 9.7) [27].

Both Meteor Crater and the Sedan Crater have a similar simple, bowl-shaped structure and so are generally referred to as "simple" craters [28,29]. Another characteristic that Meteor Crater shares with nuclear explosion craters is an elevated rim, in which the upper layers of rock have been folded back on themselves [24]. In addition, some of the material that was ejected when the craters formed falls back, either into the crater, partially filling it, or close-by, so that this also contributes to the raised rim (Figure 9.8). With increasing diameter, craters both on the Earth and elsewhere in the Solar System show the development of a more complicated structure, and not

FIGURE 9.9 Tycho impact crater on the Moon, as imaged by the Lunar Reconnaissance Orbiter. Tycho is a young lunar crater with an estimated formation age of 108 million years before present. It is located in the southern Lunar Highlands and has a diameter of 82 km, and a depth from rim to floor of 4.7 km. Unlike Meteor Crater, Tycho is a complex crater with a prominent central peak that rises 2 km above the crater floor. The edge of the crater is terraced, but the floor itself is relatively flat [30]. (Image: NASA Goddard/Arizona State University.)

surprisingly, are termed "complex" craters (Figures 9.8 and 9.9) [28,29]. Complex craters typically have flat floors, a central uplift peak, terraced margins, and show the development of extensive impact melt sheets [28,29]. Very large impact structures on the Moon and elsewhere in the Solar System, often have concentric ring structures and are referred to as impact basins rather than craters. An impressive example is the Orientale impact basin (Figure 9.10).

Finally, let's return to Meteor Crater. In the second half of the 1960s, with its extraterrestrial origin no longer in doubt, the crater was used for Apollo astronaut training (Figure 9.11). Well, as the best-preserved impact crater on Earth, where else would they have gone? There is some classic footage available of astronauts carrying out instrument field tests at the crater in 1965 [32]. Meteor Crater is still owned by the Barringer family and there is a fascinating museum on the crater rim that tells

FIGURE 9.10 The Orientale impact basin on the Moon was formed 3.8 billion years ago, at the end of the epoch known as the Late Heavy Bombardment (Chapter 4). It has a diameter of 930 km and displays a multi-ringed structure. The asteroid that formed the crater is estimated to have had a diameter of 64 km [31]. In comparison, the asteroid that wiped out the dinosaurs was only 10 km in diameter. (Image: Ernest Wright, NASA/GSFC Scientific Visualization Studio.)

the story of this extraordinary place. The Barringer family are active benefactors of the Meteoritical Society and the Barringer medal is awarded annually to the scientist who has made the most significant contribution to understanding impact processes.

In the end, Daniel Moreau Barringer was completely wrong about there being a huge iron meteorite buried under Meteor Crater. And for this error, he lost a fortune. But his obstinacy and tenacity left a lasting legacy in the form of a much better understanding of how impact events have shaped our Solar System.

FIGURE 9.11 Eugene Shoemaker describing the geology of the rim ejecta at Meteor Crater to a large group of astronauts during a field trip that took place from 16 to 19 May 1967. (Photo: United States Geological Survey.)

In an earlier chapter, we looked at the circumstances surrounding the fall of the Glatton meteorite in 1991. It would be 30 years before another meteorite fell in the UK. This new arrival was very different from Glatton. To find out more, we are off to the very picturesque Cotswold town of Winchcombe in Gloucestershire. To the driveway of the Wilcock family, to be precise.

10 The Winchcombe Meteorite

An Extraterrestrial Splat!

Now we come to the story of how an exotic space rock pitched up in a quaint corner of rural England. It had come a long way and made quite a stir.

One sunny afternoon in early March 2021, I had my very own extraterrestrial, close encounter. On that particular day, I found myself trudging up the driveway of a suburban residential home in Winchcombe, Gloucestershire (Figure 10.1) to meet Rob and Cathryn Wilcock, and their next-door neighbour. As I arrived, Rob headed indoors to fetch a plastic bag full of rocky fragments that he had found in a big heap on his driveway a few days earlier. I took one look in the bag and had a sort of "wobbly" moment. There in front of me were fragments of a rare type of meteorite that had never before been recovered in the UK. To say, I was in a state of shock doesn't

FIGURE 10.1 The beautiful medieval Cotswold town of Winchcombe, where all the action took place. (Photo: the author.)

DOI: 10.1201/9781003174868-10

really do it justice. I did my best to look professional, but I am not sure how well I suceeded. To understand why I was in a bit of a state we need to delve a little into the long history of British meteorite falls.

We have been collecting meteorites in Britain for well over 200 years. The first reliable account of a space rock landing in the UK, from which we still have material available for study, is the Wold Cottage meteorite that fell in a field in the village of Wold Newton, Yorkshire in 1795 [1,2]. It came down close to where John Shipley, a farm labourer, was working at the time and weighed in at 25 kg. It buried itself in a hole 18 inches deep. Mr Shipley must have been a little shocked by these events. It was a local sensation and a monument was ericted to commemorate its arrival. It can still be visited to this day (Figure 10.2).

Going even further back, there is a fairly reliable account of an approximately 10 kg meteorite falling near Stretchleigh, Devon in 1623. It was described as being "like a stone singed or half burnt for lime" [3]. Unfortunately, no material from the Stretchleigh stone has survived to the present day. Perhaps I should modify that statement. I am sure it has survived and likely sits in someone's garden or another unknown location, it is just not currently available for scientific study, which is a pity. If you live in that area, keep a look out. You never know! So, in the UK, we have a very long record of meteorite encounters. But in all that time, nothing like Winchcombe had ever been recovered. That was why, as the first scientist to get a sight of this amazing material, I went a little bit "wobbly", as I peered into that bag. But how had I come to walk up the drive that afternoon? Let's backtrack a little.

For me, it all began on the evening of Sunday 28 February. It was sometime between 11 and 11.30 pm, and I had been sending various data files to a colleague in the States. As I was packing up for the evening, I noticed a large number of emails from Graham Ensor, a member of the British and Irish Meteorite Society (BIMS), and a frequent visitor to the Open University where I work. Graham is a true meteorite enthusiast and seems to know about any major meteorite news event within minutes of it happening. He was emailing to tell me about a huge fireball that had just been seen all over the UK, with sightings as far as Ireland, Northern France, and the Netherlands (Figure 10.3). There was already a lot of footage of the fireball on the web. Looking at the available images and videos, it seemed to me there was a very strong likelihood that fragments had reached the surface. I sent a number of emails to colleagues involved with the UK Fireball Alliance, including Luke Daly at the University of Glasgow and Ashley King at the Natural History Museum, London. As I went to bed that evening, I had the overwhelming sense that this event was the real thing (Figure 10.3).

Over the next few days, I kept in contact with Ashley, who was co-ordinating the recovery activities. Fireball experts in France and Australia were hard at work refining the meteorite's likely trajectory and computing a potential area in which fragments might have fallen [4]. I was aware that an area close to Cheltenham in Gloucestershire was the most likely recovery zone. We had waited for 30 years since the last UK meteorite fall in Glatton, near Peterborough (Chapter 2), so this was an

FIGURE 10.2 The Wold Cottage monument. Wold Cottage was a property in Wold Newton, Yorkshire, owned by Major Edward Topham. In 1795, one of his workers, John Shipley, was in a field near the house when, after hearing sounds of thunder in the air, an object hurtled by him and hit the ground only 30 yards from where he was standing. When John Shipley located the sample, it was buried in a hole about 18 inches deep. He described it as having a black crust and smelling of sulphur, it weighed about 25 kg. The inscription on the monument reads: "*Here on this spot, December 13, 1795 fell from the atmosphere an extraordinary stone. In breadth twenty-eight inches, in length thirty-six inches and whose weight was fifty-six pounds. This column in memory of it was erected by Edward Topham, 1799*". (Photo: Graham Ensor.)

FIGURE 10.3 The Winchcombe Fireball (Sunday 28 February 2021). One of the most imaged fireball events ever. (Photo: Ben Stanley/AllSky7.)

exciting, but also a worrying timc. On the Monday following the fireball, there had been a media campaign to get the word out in the local area, asking people to keep their eyes peeled for possible meteorites [4,5]. But gradually, as I hadn't heard any positive news, I started feeling a little disappointed. I remember a telephone conversation with Graham, in which we agreed that meteorites are generally found within hours of their fall. It was not a good sign.

Computing a precise area in which fragments might have fallen was going to be a critical aspect of any successful recovery campaign [4,6]. The area in which fragments fall is called a meteorite's "strewn field" (Figure 10.4).

Over an ever-increasing area of the globe, the skies are now continuously surveyed by a network of cameras that have the capability of tracking fireballs and providing precise coordinates for their trajectories. The February 2021 UK fireball was one of the most imaged events in history [6]. The response from the public to the media campaign was phenomenal, nothing quite like it had ever been seen in the UK before. The data collected from the fireball was also used to compute a pre-atmospheric entry orbit for the object that landed in Winchcombe. It turned out that the Winchcombe meteorite had originated from the asteroid belt (Figure 10.5) [4,6].

Then on Wednesday 3 March at about 9 am, Ashley circulated an email with photos sent in by the general public living in the calculated "strewn field". It was centred on the Cotswold town of Winchcombe, Gloucestershire. I opened up the first set of images, all sent in by a single household, and had a total shock. They showed the image of a pile of dark, smashed up rocks in the shape of a big "splat" mark, with rays of debris heading in all directions. It looked very similar to other images of small

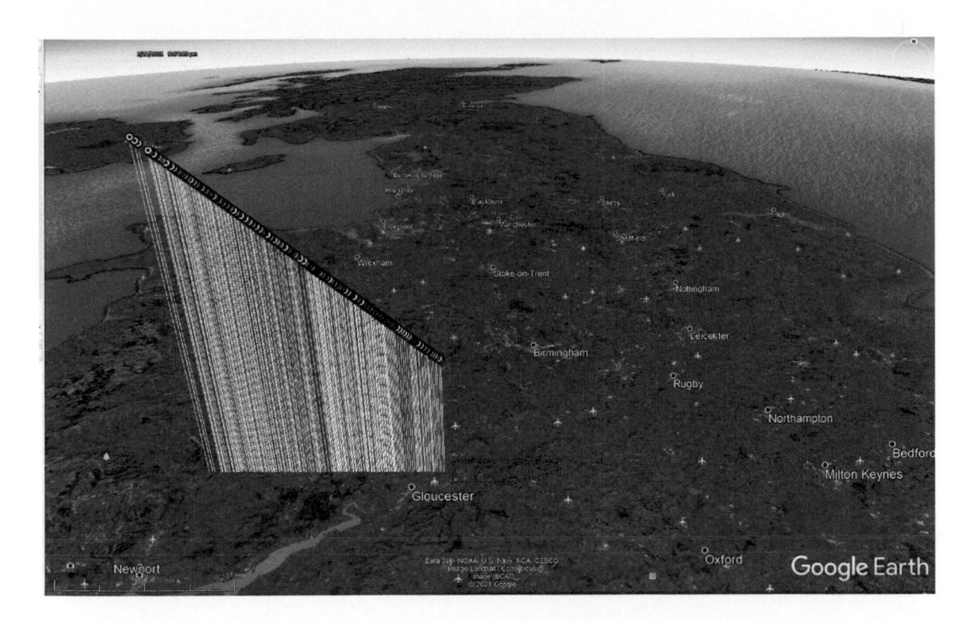

FIGURE 10.4 The vast amount of camera and image footage meant that the trajectory of the Winchcombe fireball could be calculated with a great deal of precision. This in turn meant that a limited area could be defined in which fragments were likely to have fallen. This is known as a meteorites "strewn field". (Image credit: UKMON.)

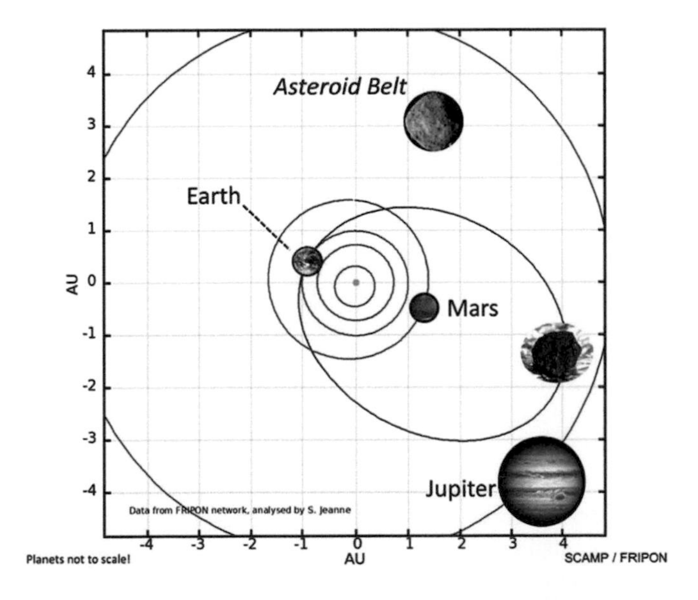

FIGURE 10.5 Most meteorites that land on Earth started out life as small objects in the asteroid belt, the rocky band of debris that lies between the orbits of Mars and Jupiter. The data collected from the fireball on Sunday 28 February confirmed that the Winchcombe meteorite also originated from the asteroid belt. (Image: Ashley King/SCAMP/FRIPON.)

FIGURE 10.6 The splat! This was the scene that greeted the Wilcock family on Monday 1 March. There was a central heap of dark debris with rays of dust and fragments. Pieces of the meteorite lay strewn across their garden and some of the bigger bits had made it onto their neighbour's driveway. Rob Wilcock carefully picked up the majority of the pieces and put them in clean plastic bags. It was only about twelve hours or so after the fall. It hadn't rained and so Winchcombe became one of the most pristine meteorites ever collected from the Earth's surface. (Photo: Rob Wilcock.)

meteorite impact structures on hard, man-made surfaces that I had seen in the past. In particular, the Braunschweig meteorite that fell in Germany in 2013 and which I had featured in my blog. Looking at the image from Winchcombe, I was instantly sure that it was a meteorite (Figure 10.6).

I immediately phoned Mahesh Anand, cajoling him out of the virtual meeting he was in. This couldn't wait. He took a look at the photos in the email and agreed, it was the real thing. Graham was unavailable for the moment, but when I finally got through to him, he was also in no doubt. We had our meteorite! Someone needed to get over there and fast. I got back to Ashley to give him the verdict of our ad hoc team. But it seemed there was a problem. We had the name of the finder, we knew he lived in Winchcombe, but we didn't have a contact number, and emails didn't seem to be getting through. I set about tracking him down. This should have been difficult, but it really wasn't. In the age of the internet, there is just so much information freely available on the web. I was able to obtain Mr Wilcock's telephone number. So, I phoned him up.

Rob Wilcock was very friendly and just a bit surprised that no one had been in contact with him earlier. There had clearly been some sort of communications glitch. He was happy to chat about his discovery and was convinced that the material deposited on his driveway was from Sunday's fireball. I said I thought so too, and we agreed to meet up in Winchcombe that afternoon at about 3 pm. I got in my car for the journey of a lifetime. This may all sound very straightforward, and in normal times, it probably would have been, but these were not normal times. We were in the middle of a pandemic with a full-blown lockdown for the whole country. Getting me on the road to Winchcombe required an enormous effort by staff based at the Open University in Milton Keynes, who had pushed through all the necessary consent and health and safety forms in an incredibly short space of time. Much of this effort was spearheaded by Professor Monica Grady, and I remain deeply indebted to her and all the staff at the Open University who worked so hard and so fast that day to make things happen. Throughout, Winchcombe was a team effort, and continues to be so.

So, there I was at 2.45 pm outside Mr and Mrs Wilcock's house. As I headed up the drive to meet them, I passed the "splat" mark. After an initial chat, Rob headed in to get the plastic bags full of sample material, and as you now know, I looked inside and had what was in fact my second "wobbly" moment of the day (Figure 10.7). So, Winchcombe was a rare type of meteorite never before recovered in the UK. Why is this important?

Starting with Wold Cottage (Figure 10.2), there had been 21 authenticated and officially classified meteorite falls recovered in the UK and Ireland, before Winchcombe turned up. Nineteen of these are examples of the most common meteorite type, known as ordinary chondrites. The other two specimens, Pontlyfni and Rowton, are respectively derived from the outer and inner portions of melted asteroids. With the exception of these two specimens, most planetary scientists the world over would agree that the combined British and Irish meteorite record was essentially as dull as dishwater. Winchcombe changed all that! For starters, it is a carbonaceous chondrite, a type of meteorite never previously recovered in the UK. Unlike ordinary chondrites, carbonaceous chondrites have a very dark interior colour. This is partially because of their relatively high carbon content, which gives this meteorite type its name. Carbonaceous chondrites typically contain a complex mix of organic molecules, including alcohols, amino acids, and a rich variety of tar-like substances [4,7]. At least some of these compounds may have originated within the molecular cloud from which the Solar System was formed, or alternatively at the cold outer edges of the Solar System itself [8,9].

FIGURE 10.7 Photo taken just after my second "wobbly" moment. Rob Wilcock (right), Cathryn Wilcock (left) with fragments of the Winchcombe meteorite, safely stored in the bags in which they had been placed on the morning of Monday 1 March 2021. (Photo: the author.)

Although Winchcombe is likely a fragment from an asteroid that currently lies in the asteroid belt (Figure 10.5), scientists have good reason to believe that such asteroids originally formed much further out in the Solar System and were then scattered into the asteroid belt [10]. As we saw in Chapter 6, Jupiter is the potential culprit here. The model known as the "Grand Tack" proposes that after its formation, Jupiter moved towards the Sun, scattering any planetesimals that were in its way. It got as far as the current location of Mars and was then pulled away from the Sun, as a result of an interaction with neighbouring Saturn. As they headed outwards, the giant planets scattered the icy worlds that they encountered along the way. Some of them would have ended up in the asteroid belt. There is no doubt, based on recent studies, that Winchcombe is a carbonaceous chondrite [4,11,12], and has a composition that indicates it formed in the outer Solar System [12]. It had a long journey to make to get to Winchcombe.

And now we come to an interesting question. As I stared into that bag of broken fragments, how come I was so convinced that this was such a rare type of meteorite?

Well, it was just like coming across an old and dear friend. Back in the 1990s, I had worked at the Natural History Museum, London (Chapter 2) on an important meteorite that fell in South Africa in 1838, called Cold Bokkeveld. It was also a carbonaceous chondrite and showed lots of similarities to the material that Rob Wilcock had collected from his driveway. Carbonaceous chondrites contain small millimetre-sized objects known as calcium-aluminium-rich (CAIs) that are used to date the

FIGURE 10.8 A fragment of the Winchcombe meteorite about 5 cm in long dimension, showing a slightly cracked fusion crust on the right-hand side, and a very dark interior with white flecks on the left-hand side. Those white bits are silicate-rich objects including chondrules and calcium-aluminium-rich inclusions (CAIs), the oldest known Solar System objects, dated at 4,567 Myr [13]. Winchcombe is a carbonaceous chondrite and so is rich in organic material, including amino acids, tar-like substances, and even alcohol. (Photo: the author.)

formation of the Solar System at 4,567 million years [13]. I could see some of them in the Winchcombe fragments in that bag (Figure 10.8).

There were also chondrules, which to be fair are found in a wide range of meteorites. They are tiny glassy spheres of material that formed as molten droplets during the dusty early days of Solar System history, long before planets like Earth had formed. Carbonaceous chondrites even contain grains that are older than our Solar System, called presolar grains. But they are too tiny to see with the naked eye. The implications of what I had in front of me were enormous. After 30 years, we finally had another British fall, and not just any fall, a carbonaceous chondrite. It was a lot to take in.

After further discussions with Rob, Cathryn, and their neighbour. I phoned various members of the scientific team, including Ashley King, Monica Grady, and Mahesh Anand. Everyone I spoke to was extremely excited by the news. But what to do next? It was absolutely clear to me that Ashley from the Museum needed to come over as quickly as possible, before word of this got out beyond members of the scientific community. The Hyde, where the material had been recovered, was a residential area, and this was certainly not an event that would stay a secret for very much longer. Ashley didn't need much persuading and was soon on the next available train from London to Cheltenham. He then took a taxi to Winchcombe. We met outside the beautiful St. Peter's church in the heart of medieval Winchcombe. It was already dark by the time Ashley's taxi drew up. Before heading back to meet up with Rob and Cathryn, there was a period of intense discussion over the phone with a range of members of the UK meteorite community. Following these excited conversations, we headed back to the Wilcock's house to go through all the samples in detail.

Due to the pandemic, we all had to sit in the garden. It was now very dark, and we must have looked a very strange sight. Illuminated only by the lights shining out from the lounge, we sat around with the samples laid out on a garden table. It was an amazing couple of hours, as we studied the various carbonaceous chondrite fragments. We were joined by Hannah Wilcock, Rob, and Cathryn's daughter, who on the Sunday evening when the meteorite had fallen recalled hearing a "shattering sound, as if a photo frame had fallen and broken". She hadn't gone out to look and see what the source of the noise might have been, as it was so dark. We discussed chondrules, CAIs, and presolar grains with our hosts, who had in the intervening period since my first visit been doing a bit of internet searching for information about meteorites. Ashley concurred with my initial suggestion that the Winchcombe material was closely similar to Cold Bokkeveld. With the agreement of the Wilcock family, Ashley took the material away with him to be curated at the Natural History Museum. As we finally headed off, I picked out two small fragments of Winchcombe to get oxygen isotope analysis underway at the Open University as quickly as possible [14]. I then had to head home in a hurry, as I had an important late-night meeting with colleagues in Japan about the Hayabusa2 asteroid sample return mission. It was very foggy and I almost didn't make it on time.

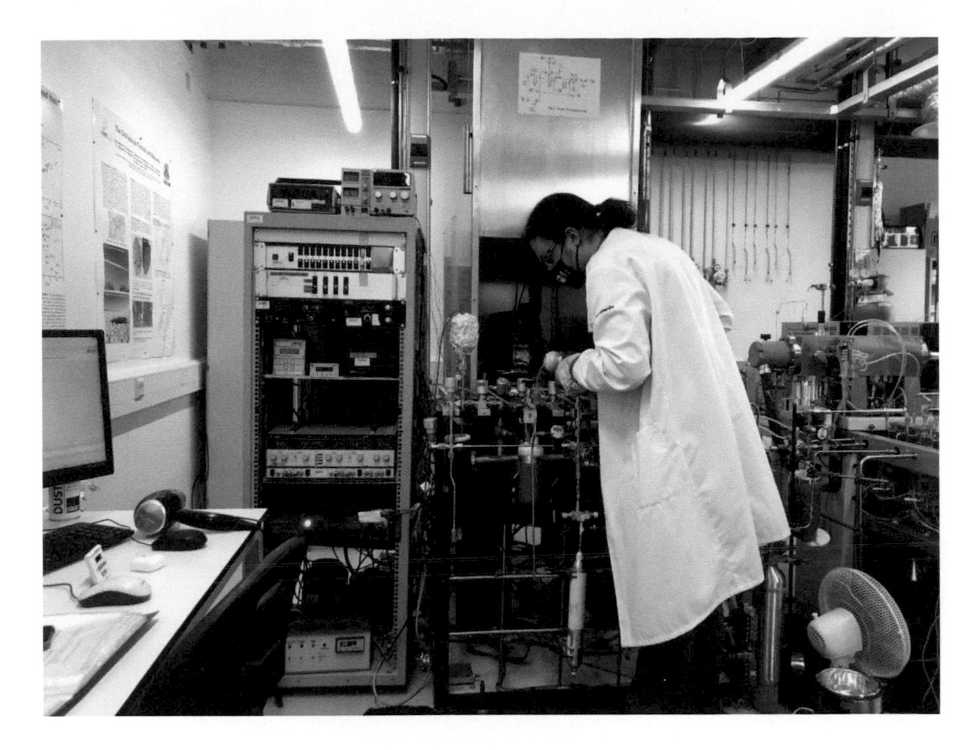

FIGURE 10.9 Ross Findlay at the Open University running a sample of Winchcombe to analyse its oxygen isotope composition. The results of this analysis confirmed that Winchcombe was a carbonaceous chondrite. (Photo: Vincent Deguin.)

The next day, I headed to Buckingham for a meeting in the Tesco car park with Mahesh. I handed over the two small Winchcombe samples and he then took them to the Open University, where Ross Findlay was ready and primed to carry out the oxygen isotope analysis. Why was it important to get this done quickly? Well, as we have seen, oxygen isotope analysis is a particularly powerful technique that can help to identify which group an individual meteorite belongs to [14]. This analysis confirmed that Winchcombe was indeed closely related to the historic Cold Bokkeveld meteorite. Both are members of the CM group of carbonaceous chondrites (Appendix 2). We all felt that we had achieved the world record for the fastest oxygen isotope analysis of a newly fallen meteorite: from arrival on Sunday evening, to analysis on Friday morning, had taken less than five days. (Figure 10.9).

The following day (Friday 5 March), I headed back to Winchcombe, via the village of Woodmancote. My job this time was to visit Mr and Mrs Carrick to check out the stone they had found in their garden. All I had to go on was their postcode. Again, there had been a bit of a communications glitch. The postcode wasn't so helpful, it just narrowed things down to a long suburban street. And so, I had to do a spot of cold calling, "I'm looking for Mr and Mrs. Carrick", "Never heard of them!"

FIGURE 10.10 The Winchcombe sample found in Mr and Mrs Carrick's garden has a very dark interior. The white flecks are CAIs and chondrule fragments. It shows a complex exterior with what appears to be, two types of fusion crust. The orange-coloured crust may be slightly older than the dark black fusion crust, which seems to spill over one edge of the fragment. The specimen is about 3 cm in long dimension. (Photo: the author.)

It took a while! But finally, I found someone who knew them and sent me in the right direction. They turned out to be lovely people who were only too happy to donate the stone they had found to science. It was small and dark, with what appeared to be two distinct generations of fusion crust on the outside. Mr Carrick knew it was newly fallen because he was actively surveying his garden due to an intruder cat that was being a bit of a nuisance. Later that day, I joined one of the searches that was taking place across the open farmland surrounding Winchcombe. We didn't find anything that day. But the next day, a team from the University of Glasgow led by Luke Daly, did locate a superb stone while searching the local fields [4] (Figure 10.10).

More small fragments of Winchcombe were located in the months that followed the fall. The scientific effort also swung into motion. A "consortium" study was set up to investigate all the important aspects of the Winchcombe event. An overview paper was published first [4], and then a series of more specialist papers, as part of a volume dedicated to Winchcombe [5–7,11,12].

FIGURE 10.11 In the days following the fall of the Winchcombe meteorite, teams conducted slow painstaking searches of the surrounding countryside. This led to the discovery of a significant new fragment by a team from the University of Glasgow. (Photo: the author.)

Following the fall of Glatton in 1991, we waited a long time for a new meteorite to arrive in the UK. But it was worth it, Winchcombe is a very special meteorite. But now it's time to up the tempo and travel back to 1969. The year everything happened, more or less, all at once (Figure 10.11).

11 1969 and All That

As wine lovers know only too well, you have good years and bad years. But from time to time, you can have a truly exceptional one. 1969 was that sort of year for scientists studying space rocks. Two important meteorites hit the Earth. Moon rocks were brought back by the Apollo 11 astronauts and meteorites were found in Antarctica. Nothing would ever be the same again. Here's why.

Sometimes stuff just happens! You go for long periods bouncing along the bottom, usual routine, all very mundane, nothing to write home about. A bit boring perhaps. And then it all goes a bit nuts. That's life really! It's the same with years. I bet 1065 was a pretty dull year in England. Not much going on. People doing their regular day-to-day jobs. Lots of chat down the local tavern about the price of eggs, the weather, who would win the largest onion competition at the village fete. You know the sort of thing. And then along comes 1066, two invasions. One in the north, one in the south. Two big battles, the fist one, a total win for the Anglo-Saxons against the Norwegians at Stamford Bridge [1] and the second, a total wipe-out against the Normans at Hastings [2]. And that was that. World turned upside down. And there have been other years like that since: 1745, the year Bonnie Prince Charlie almost retook the English throne back for the Stuarts; 1776, the year the American colonies decided it was time to get serious about independence; 1815, the Battle of Waterloo; 1914 and 1939, the start of two world wars; and 1966, the last and only year England won the World Cup.

And finally, 1969. It turned out to the greatest year ever for extraterrestrial sample analysis. Four major events that year completely changed our understanding of the Solar System and our place in it. 1969 can be summed up in four words: Allende, Apollo, Murchison, and Yamato. Let's look at each in chronological order and see what all the fuss was about.

There never was, and may never be again, an event like the fall of the Allende meteorite. To appreciate why it was so significant, we need to travel back in time to Christmas 1968. On 21 December, Apollo 8 blasted off from the Kennedy Space Centre becoming the first manned spacecraft to leave Earth orbit and the first to reach the Moon. Apollo 8 made ten circuits of the Moon and then returned safely to Earth, splashing down in the Pacific on the afternoon of 27 December 1968. It was a mission that gripped the public imagination in a unique way. The "Earthrise"

DOI: 10.1201/9781003174868-11

photograph taken by the Apollo 8 astronauts had a profound effect on mankind's perception of its place in the universe (Figure 11.1). It showed Earth as a colourful jewel in a vast sea of darkness. The reading of the first ten verses of the Book of Genesis by Apollo 8 astronauts during a live broadcast from the spacecraft on Christmas Eve 1968 was an emotional event that was watched by a huge television audience. Apollo 8 was a technical and scientific triumph and made it possible for the Moon landings to take place the following year [3]. Any lingering doubts about the feasibility of launching a successful mission to land on the Moon in 1969 were ended by Apollo 8.

As 1969 dawned, the scientific community was going into overdrive in preparation for the anticipated return of lunar samples following a successful Moon landing later that year. There was now no doubt about it, NASA would be shooting for a lunar landing mission before the end of 1969. This was in fulfilment of President John F. Kennedy's historic promise made to a joint session of Congress on 25 May 1961 that the US "*should commit itself to achieving the goal, before the decade is out, of landing a man on the Moon and returning him safely to the Earth*".

But so much to do and so little time left to do it in. Activity in the labs that were due to receive samples of Moon rock was frenetic. New techniques were being developed, instruments fine-tuned, and test samples analysed. All sorts of new protocols

FIGURE 11.1 "Earthrise" – A view of the Earth from the capsule of Apollo 8 taken by astronaut Bill Anders. (Image: NASA/Bill Anders.)

and procedures were being devised. There had never been anything quite like it. And what would these samples look like? How old were they? What sort of composition did they have? No one had a clue. After all, no one had ever been to the Moon before. How could you check out your instrumentation when you didn't know what you were going to be analysing?

And then it happened!

In the very early hours of 8 February 1969 (1.05 Central Standard Time), a huge fireball lit up the skies over a large area of northern Mexico and adjoining parts of the US [4–6]. The path of the fireball was from southwest to northeast, and as is usual for such large meteoroids, it exploded violently as it penetrated to lower altitudes in the atmosphere. This resulted in the formation of thousands of individual fragments which landed over a huge area measuring approximately 8 by 50 km (Figures 11.2). One relatively large stone weighing about 15 kg fell within 4 m of a house in the village of Pueblito de Allende, Mexico. As a consequence, the meteorite was named Allende.

Less than two days after its arrival, Dr Elbert King (Figure 11.3) from NASA's Manned Spacecraft Center in Houston, Texas visited the area and collected nearly 7 kg of material. Dr King went on to write the first scientific paper about Allende which was published in the prestigious *Science* journal within 20 days of the meteorite's fall [7]. News of the Allende's arrival spread rapidly within the scientific community, and on 12 February 1969, Roy Clarke Jr. and Brian Mason from the Smithsonian Institution in Washington DC begin working in the area, collecting

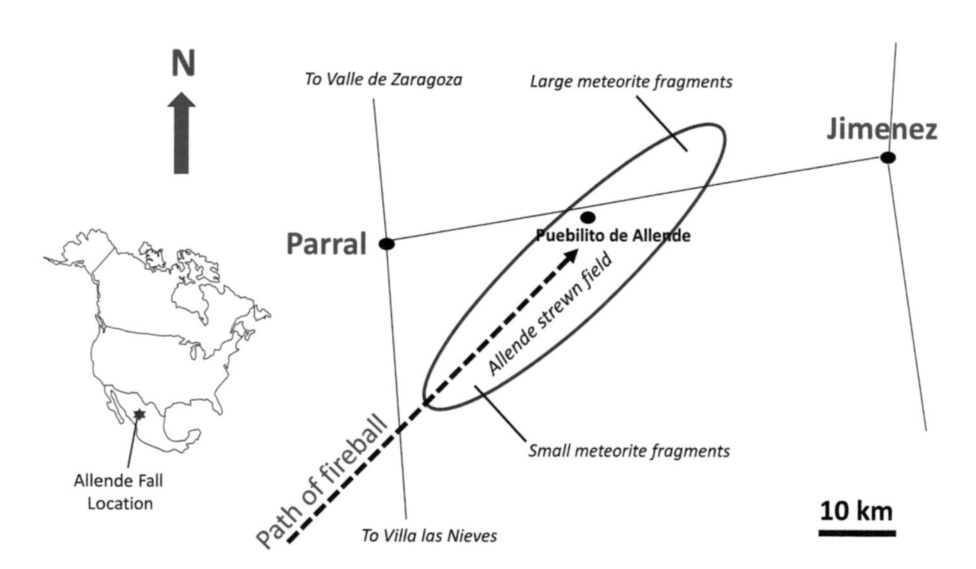

FIGURE 11.2 The breakup of a large meteorite as it penetrates into the lower atmosphere is a common occurrence. The Allende fireball was no exception. Following such a fragmentation event, stones from the fall are distributed over an elliptically shaped area that is known as a strewn field. As is commonly the case, the largest fragments of Allende travelled the furthest after the breakup event due to their greater momentum. This resulted in a general decrease in the size of the fragments from northeast to southwest [5]. (Image: the author.)

FIGURE 11.3 Dr Elbert King (left) showing a visitor around part of the Lunar Receiving Laboratory (LRL), where the Apollo 11 samples were initially stored and processed. The Allende meteorite samples were collected by Dr King with the principal motive of using them to test some of the systems and facilities within the LRL. (Photo: NASA.)

samples and documenting details of the fall (Figure 11.4). Their account of the Allende Fall and the follow-up field studies remains the most comprehensive description of the event [4]. The Smithsonian team made multiple trips to the area, particularly during November 1969, and collected more than 1,000 stones weighing in excess of 400 kg [5].

I once had the privilege of visiting the meteorite storeroom in the basement of the Smithsonian Museum. It is an absolute treasure house of extraterrestrial marvels. But my lasting memory of the visit was Allende. It was just the sheer number of specimens of this incredible meteorite that was so amazing. I had seen lots of Allende samples before, we even have Allende material in our collection at the Open University. But to see row upon row of dark, partially to fully crusted, samples of such a beautiful meteorite was an emotional and unforgettable experience. Like Winchcombe (Chapter 10), Allende is a carbonaceous chondrite, but the two are in fact very different. We will look at this in more detail in a bit.

The Smithsonian Institution was not the only scientific organisation that actively collected stones from the Allende strewn field. Arizona State University and the Field Museum of Chicago also obtained large quantities of material. There was also a significant amount of activity from private dealers and collectors. Local people played a major role in the collection effort (Figure 11.5). No one really knows how much Allende was collected. The official estimate is in excess of 2 tons [8]. To put this in context, the next largest carbonaceous chondrite by mass is the Kainsaz

FIGURE 11.4 Brian Mason of the Smithsonian Institution (centre) holding a 12 kg Allende stone that he had recovered on 17 February 1969. Other individuals in the photo are unidentified [5]. (Photo courtesy of Smithsonian Institution.)

FIGURE 11.5 Local school children and their teacher collect pieces of the Allende meteorite following its fall on 8 February 1969 in Mexico. (Photo: Smithsonian Institutional Archive.)

meteorite that fell in Russia in 1937 and that weighed in at just 200 kg, ten times less than Allende. This was a case of serendipity on steroids. All over the world, scientific labs were in the final stages of preparation for the return of Moon rocks and suddenly here was this fresh extraterrestrial material in abundance, which could be used to calibrate their instruments. In particular, Elbert King was keen to use Allende to test protocols at the Lunar Receiving Laboratory Radiation Counting Facility (Figure 11.3). The reason for this was that as an airless body, the surface of the Moon is continually bombarded by high energy particles, both from the Sun and elsewhere in the galaxy. These high energy interactions induce radioactivity in Moon rocks. Before arriving on Earth, Allende would have undergone the same type of irradiation. As a result, Allende material could be used to test the laboratory techniques that would then be put to work on the returned lunar samples. And there was more than enough for everyone.

And Allende had one more surprise, a big one. On the surface of almost every stone, where the dark fusion crust had fallen away, the interior structure of the meteorite could be seen very clearly. And there, in plain sight, were these light-coloured centimetre-sized, irregularly shaped "things". For want of a better word, they were called "inclusions", probably because it sounds more sciencey than "things". When they were analysed, they turned out to be enriched in calcium and aluminium compared to the bulk of the material making up the meteorite. Scientists, not being very poetic types, called them calcium-aluminium-rich inclusions, and as that was a bit of a mouthful, it got shortened to CAIs (Figure 11.6). The name has stuck, even though it does sound a bit like the US government's spy agency. But what are they? Allende had lots of big chondrules, rounded droplets that may have formed as fiery rain.

FIGURE 11.6 Close up of an Allende specimen showing a group of four white CAIs. Each is a little under 1 cm in diameter. They have very irregular shapes in comparison to chondrules, which are little bit like cosmic footballs – especially the one on the left. (Image: Andy Tindle.)

But the CAIs didn't look anything like chondrules (Figure 11.6). And their mineral compositions were also completely different. It was all a bit of a mystery.

By coincidence, these objects had already, and very recently, been described in a different meteorite [9]. And there was some theoretical work to suggest that these were very important objects indeed [10]. When stars form, things get very hot. In this intense heat, solid dust is rapidly vapourised. But what happens if some of this vapourised material escapes from the stellar furnace? Well, it will turn back into a solid again, a process known as condensation. Calculations had been made by scientists in the past about what the "condensates" formed during such a process would look like. But it was Lawrence Grossman, then at Yale University, who pointed out that the Allende CAIs looked a lot like condensates from such a hot gas, or at least the high-temperature minerals that would form as such a gas cooled. And consistent with this possibility, CAIs turned out to be very old [11]. The most accurate dating suggests that CAIs are the oldest solid objects that formed in our Solar System. The dates that have been obtained indicate that they formed 4,567 million years ago [11,12].

But what sort of environment would CAIs have formed in?

Astronomical observations show that as dust and gas is swept up by a young growing star, things can get extremely energetic. T-Tauri stars are young, roughly Sun-sized objects [13] that show violent convulsions, with massive discharges of material, particularly from their polar regions, generally referred to as bipolar outflows (Figure 11.7) [12]. It is likely that the Sun went through a similar high energy phase early in its evolution. It has been suggested that CAIs are objects that were formed during the Sun's T-Tauri phase [14]. In this scenario, CAIs would be the products of gaseous material that was ejected from the growing Sun, condensed into solid objects, and then mixed with surrounding cooler dust. Later, this mix of dust and CAIs would have coalesced to form the asteroids, one of which would have been the source of the Allende meteorite.

If the only significance of Allende had been to provide abundant test samples for Apollo mission-related activities, it would almost certainly have fallen into obscurity long ago. But Allende genuinely changed our understanding of how the Solar System formed. It ushered in a revolution in ideas about early Solar System processes. And like most revolutions, things got a bit out of hand. But we will get to that in Chapter 16.

The next big event in 1969 was the Apollo 11 Moon landing. Launched on 16 July 1969, it was without doubt the most important manned space mission of all time. On 21 July 1969, Neil Armstrong became the first person to walk on the Moon and was followed out onto the lunar surface 19 minutes later by Buzz Aldrin (Figure 11.8). Their activities on the surface of the Moon only lasted just over two hours, but it was without doubt a moment of lasting and major importance in the history of humanity. The Apollo 11 astronauts collected and returned to Earth 22 kg of lunar samples. Subsequent analysis of these precious materials provided important insights, not only relevant to the formation and evolution of the Moon, but also about the nature of the solar wind and how it interacts with the Moon's surface layers.

As we have seen, 1969 was already a vintage year for space science and then on 28 September, it got even better. At just before 11 am local time, a large fireball that was moving from the southeast to the northwest exploded over the township

FIGURE 11.7 Young stars are highly energetic objects. Not only is material falling onto the star and so making it grow larger, but in addition material is being lost during violent energetic outbursts. In this image a young, energetic star, hidden by gas and dust, is throwing out huge, elongate jets of ionised gas, called bipolar outflows. CAIs may have formed at an early stage in the Sun's development when it was going through a T-Tauri phase. (Image: NASA/ESA.)

of Murchison, Australia. Murchison is located some 140 km to the north of Melbourne in the province of Victoria (Figure 11.9) [15,16]. Reports suggest that the fireball was bright orange to red in colour with a silver rim and that it left a blue smoke trail that took several minutes to disperse [15]. A sonic boom accompanied the arrival of the fireball, rattling doors, and windows, but otherwise causing little significant damage. Debris from the fireball was scattered over an area of at least 11×3 km.

Efforts to collect up pieces of the meteorite were relatively restrained due to worries about meeting up with a less than friendly local inhabitant, tiger snakes! [16]. This is a reasonable concern as estimates suggest that there is a 40%–60% mortality rate from untreated tiger snake bites!

The official record for Murchison indicates that a very respectable 100 kg of material has been recovered from the strewn field [17]. Pieces of the meteorite were still being found as late as 1990 [16]. The most successful team searching for pieces of Murchison were two brothers, Peter and Kim Gillick, aged ten and eleven at the time [18]. Between them, they collected about one-third of the total recovered mass of Murchison (Figure 11.9).

FIGURE 11.8 The Apollo 11 Moon landing in July 1969 was a major technological achievement and a scientific event of lasting importance. Analysis of the lunar samples returned to Earth by the mission resulted in fundamental advances in space science. Left – Buzz Aldrin salutes the US flag. Right – Apollo 11 lunar module, the Moon and Earth (Photos: NASA.)

FIGURE 11.9 A view of main street Murchison, Victoria, Australia where on 28 September 1969 one of the world's most important meteorites showed up for 1969 party! (Photo: Mattinbgn/Wikipedia.) Inset: Kim Gillick and his mother Emily with pieces from their Murchison collection. (Photo: Gillick Family.)

Of course, Murchison arrived late to the party. The world's space laboratories already had 2 tons of Allende, with its amazing CAIs, to keep them busy. And then there was the small matter of 22 kg of Moon rock. What more could Murchison add to this extraordinary inventory. As it turned out, quite a lot, because Murchison is a very different sort of meteorite to Allende. True, they are both carbonaceous chondrites, but Allende is a coarser-grained CV3 type (C stands for carbonaceous, V is for Vigarano, the meteorite that defines the group, and 3 just says Allende was largely unheated when it was inside its source asteroid; see Appendix 2 for further details). Murchison on the other hand, like Winchcombe (Chapter 10), is a member of the CM2 group (again C stands for carbonaceous, M is for Mighei, the type meteorite of this group, and 2 signifies that it was altered by heated fluids within its source asteroid). That last bit might sound somewhat strange, but CM2s show clear evidence for the action of heated fluids (mainly water) that would have been present in their source asteroids very early in Solar System history [19]. When these asteroids first formed, they almost certainly contained a significant amount of ice. But this soon melted as the asteroid began to heat up, mainly due to the decay of short-lived radioactive species, in particular, an unstable form of aluminium ^{26}Al [20]. These heated fluids changed the original minerals in the CM2 meteorites to water-bearing clays and various other related hydrated minerals [19].

Water-bearing CM2 meteorites like Murchison are in fact the most common variety of carbonaceous chondrite. To date, there have been 21 officially recognised CM2 falls, representing 40.4% of all carbonaceous chondrite falls. In comparison, there have been only seven CV3 (Allende-like) falls, which is just 13.5% of all carbonaceous chondrite falls. But why does this matter? Actually, it may be the reason that Earth is a habitable planet. Life as we know it may only have been possible because of the relative abundance of water-bearing meteorites like Murchison. The Earth formed in the dry inner Solar System and was probably a barren sterile world when it first reached roughly its present size. Something was needed to bring in the water from the outer regions of the Solar System. The suggestion is that water-bearing, Murchison-like, meteorites hydrated the early Earth, making it a fit environment for the evolution of life (Figure 11.10) [21].

But transporting water to early Earth may have been only one part of the story. Scientists studying the newly arrived Murchison meteorite began to look at the inventory of organic material that it contained. It was quickly realised that Murchison, like other CM2s, contains an extraordinary diversity of organic molecules [22–25]. These range from relatively simple compounds such as alcohols, amino acids, aldehydes, and ketones to very complex, so-called "macromolecular" material, basically a thick tar-like substance [23,25]. Naturally, attention has been focussed on those molecules, such as amino acids, that could have been important to the evolution of life on Earth. It was quickly established that the amino acids in Murchison could not have been biological in origin. No alien lifeforms were killed in the making of the meteorite. But why such certainty that these molecules were synthesised purely by chemical processes? This is all down to an important feature of many organic molecules termed chirality (Figure 11.11). Here's how it works.

When you look at your hands, they seem to be identical, with both having four fingers and a thumb (Figure 11.11). But appearances are deceptive. Put one hand on

FIGURE 11.10 Early in Solar System history, asteroids similar in composition to Murchison may have transported water and organic compounds to Earth making it a habitable planet. (Images: NASA.)

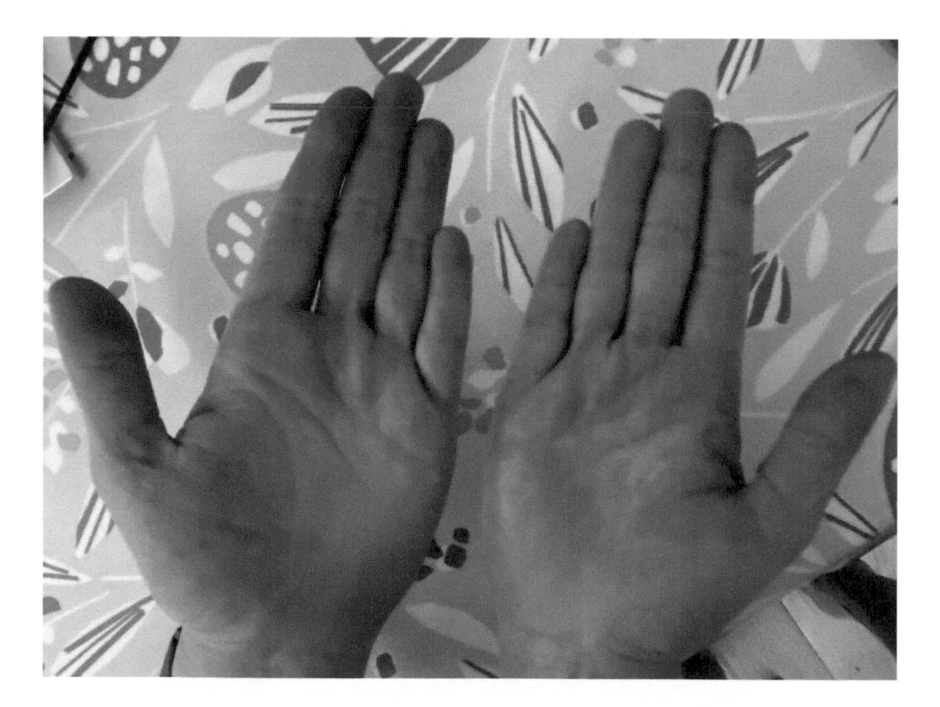

FIGURE 11.11 Your hands look identical, but they are not. They are in fact mirror images of each other. Hold both on top of each other palms down and you will see that they are non-superimposable. Many molecules have the same property and are said to be chiral. See text for further discussion. (Photo: the author.)

FIGURE 11.12 Like your hands (Figure 11.11), these molecules are mirror images and non-superimposable. Biology selects for just one of these two forms, whereas chemistry goes for a 50:50 mixture of both. Using this evidence, it is clear that the amino acids and other organic molecules in the Murchison meteorite were formed by chemistry not biology. (Photo: the author.)

top of the other with both palms facing down and it quickly becomes clear that one hand is arranged in the opposite sense to the other. Technically, your hands are said to be "non-superimposable". If you now hold them, palm facing palm, and imagine that a mirror runs between them, something even stranger is apparent. Your hands are like reflections of each other in a mirror. Many molecules show the same characteristics. They come in two forms that look the same but are in fact mirror images of each other and are also "non-superimposable". The two molecules in such a pair are called enantiomers (sorry about that) (Figure 11.12). Just as we talk about handedness in humans, so with molecules. One of the two mirror-image molecules is said to be the "left" enantiomers and the other is the "right". And now we get to the point of all this. If these handed molecules are produced by a purely chemical reaction, then both the left and the right forms will be present in equal amounts. But this is not the case for biology. Life only wants one of the two enantiomers. Depending on the molecule concerned, it could be left or it could be right. But only one is produced by biological processes. In the case of life on Earth, almost all amino acids are left-handed. In contrast, sugars produced by living organisms are almost always right-handed. So, when the Murchison amino acids were analysed, they turned out to be very close to a 50:50 mixture of the left- and right-handed forms. This strongly suggested that they were produced by chemistry not biology (Figure 11.12).

But there is a twist to this tale. Why did biology choose the left-handed form for amino acids in the first place? It is a bit like the question of why your country drives on the side of the road that it does. Once the decision is taken, there is really no going back. But how and why was this decision taken in the first place? Was it the toss of a

FIGURE 11.13 Emperor penguins in Antarctica. The coldest place on Earth produced the final extraterrestrial surprise of 1969. (Photo: Christopher Michel/Wikimedia Commons.)

coin? Unlikely! In the same way, why did biology decide to just use the left-handed form of amino acids? It remains a mystery. Interestingly, there is now evidence that some amino acids in Murchison and other CMs are not totally a 50:50 mixture of the two enantiomers. There is sometimes a very slight excess of the left-handed amino acid enantiomer. Not enough to invoke biology, but enough to hint at the possibility that organic molecules delivered by meteorites may have pushed life to go left instead of right when it comes to amino acids [26].

With Allende, Apollo, and Murchison in the bag, surely that was it. 1969 could not provide any further space-related surprises? Well, it could, and it did! In December 1969, Japanese scientists working in the Yamato mountains of Antarctic discovered nine meteorites. In itself that might not seem like a lot, but the implications of this discovery were profound. The Japanese group were able to show that some sort of concentration mechanism was operating, increasing the number of meteorite samples present in certain relatively well-defined areas of the Antarctic glaciers known as blue ice fields. As a direct result of those nine meteorites found in the Yamato mountains in 1969, Antarctica has become the largest and most important place on Earth for collecting space rocks. We need to find out more about all of this. It's time to head south to the frozen wastes of Antarctica (Figure 11.13). Remember to pack your thermals!

12 Antarctica
Cool Rocks from a Cool Place

We all know that the fridge is the best place to keep stuff fresh. But who would have thought that the world's largest deep freezer was doing the same job for a bunch of space rocks. We are off to Antarctica – the coolest place on the planet for meteorite hunting.

It is an astonishing fact that out of just over 73,000 officially approved meteorites logged on the Meteoritical Bulletin database (February 2024), close to 44,000 were collected in Antarctica. That means that 60% of the world's classified meteorites come from the most remote and inhospitable place on Earth. You won't be surprised to learn that not a single one of these frozen extraterrestrial samples is a "fall", i.e., a sample that was collected following its observed passage through the Earth's atmosphere. They are all "finds", collected by well-organised national and international scientific expeditions (Figure 12.1). While collection activities were seriously curtailed by the pandemic, the US ANSMET (ANtarctic Search for METeorites) programe, the biggest player in the field, was able to return to active meteorite searching during the 2023–2024 Antarctic field season [1]. As we shall see, there are very good scientific reasons why meteorite hunting in Antarctica is likely to continue for the foreseeable future. It's time to fly south and visit the greatest extraterrestrial treasure trove on our planet – Antarctica.

This remarkable scientific endeavour really got started in December 1969, when nine meteorites were located in the Yamato Mountains by members of the tenth Japanese Antarctic Research Expedition (JARE-10) [2] (Figure 12.2). A further small batch of meteorites was located in the same area in 1973. These discoveries led to a more systematic search in 1974 by JARE-15, which collected a further 663 samples. In fact, the Japanese finds were not the first meteorites to be found in Antarctica. A meteorite had been discovered during the Douglas Mawson Australasian Antarctic expedition (1911–1914). It was a 1 kg ordinary chondrite officially known as Adelie Land [3,4]. In the following years up to 1969, only three more meteorites were recovered in Antarctica [5]. As Derek Sears pointed out in his 1979 New Scientist article, until the Japanese discovery, everyone thought Antarctica was the worst place in the world to locate meteorites: "*it seemed that the climate and low population of*

DOI: 10.1201/9781003174868-12

FIGURE 12.1 Recovering a meteorite in Antarctica. Dr Romain Tartese contemplates a newly discovered meteorite in the snows of Antarctica. This image sums up why the frozen continent is such a great place to hunt for meteorites. Away from any source of terrestrial rocks, a single dark space rock stands out clearly against the light-coloured background of ice and snow. However, it is not always quite so straightforward. Meteorites in Antarctica are often found within areas that contain abundant terrestrial rock fragments. (Photo: Katherine Joy/Lost Meteorites of Antarctica.)

Antarctica meant it was the most unlikely place on Earth to expect to find them" [5]. The Japanese Yamato discoveries changed everything. Firstly, because they demonstrated that meteorites were likely present in Antarctica in large numbers, raising the possibility that systematic searches in the frozen south would significantly increase the number of extraterrestrial rocks available to science. And secondly, they provided an insight into why there seemed to be so many meteorites in Antarctica. We will get to that in a moment.

The nine Yamato meteorites from that first field season have the official names Yamato 691 to 699 (Figure 12.2). Amongst these samples are at least four distinct meteorite types, an enstatite chondrite (Y-691), diogenite (Y-692), carbonaceous chondrite of the CK group (Y-693), and ordinary chondrites of the H group (Y-694 to Y-699). This means that they are not just bits from the same meteorite, but instead were deposited during four different events. This was also consistent with Antarctica being a good place to look for meteorites.

The importance of the Japanese Antarctic meteorite discoveries made a big impact on one US meteorite scientist, the great William "Bill" Cassidy [6–8] (Figure 12.3). He put in a series of funding requests to the US Antarctic Research Programme to

FIGURE 12.2 The samples that started it all. On the left is Y-691, an enstatite chondrite, and the first sample collected by the tenth Japanese Antarctic Research Expedition (JARE-10) in 1969. On the right are all nine specimens from that first successful expedition, which demonstrated that Antarctica was the best place in the world to find meteorites. In the years since JARE-10, the number of meteorites collected from this natural freezer has grown to just over 44,000 officially classified specimens (February 2024). (Photo: the author.)

FIGURE 12.3 William "Bill" Cassidy collecting a meteorite in Antarctica during one of the early ANSMET field seasons. Inspired by the success of Japanese Antarctic meteorite collection activities, Bill Cassidy obtained the funding to set up the US ANSMET programme, with the first expedition in 1976–1977. ANSMET has gone on to be the most successful collection programme in Antarctica. (Photo: NASA/ANSMET/# S80–35604.)

send expeditions down to Antarctica to hunt for meteorites [6]. Eventually, he succeeded, and the first US expedition took place in the Antarctic summer of 1976–1977. The first three ANSMET expeditions were in fact run jointly with the Japanese. During the first ANSMET season, nine specimens were recovered from the Allan Hills, an area that lies close to the giant McMurdo research base. The ANSMET programme [9] has been going continuously ever since, although as discussed above, field parties have not always been able to make it into the field or have been constrained by various logistical issues. And if success was just about numbers, which it isn't, the results of this activity are impressive. Up to the end of the 2019–2020 field season, ANSMET had collected 23,400 samples [9]. The Japanese National Institute of Polar Research (NIPR) meteorite collection programme also has some very impressive statistics, with over 17,000 specimens recovered to date [10].

But no matter how impressive these statistics are, it is important to acknowledge the significance of the first Yamato find. As Bill Cassidy makes it clear in his book [7], there was a lot of luck involved in that initial discovery:

> One of them, Renji Naruse, picked up a lone rock that was lying on the vast bare ice surface and recognized it as a meteorite. In the preceding 200 years only about 2000 different meteorites had been recovered over the entire land surface of the earth, and finding a meteorite by chance must be counted as extremely improbable.
>
> It's lucky, therefore, that this initial discovery at the Yamato Mountains was made by a glaciologist, who would not be expected to have a quantitative understanding of exactly how rare meteorites really are, and of what a lucky find this should have been; Naruse and his companions proceeded to search for more. By day's end they had found eight more specimens in a 5 × 10 km area of ice – a tiny, tiny fraction of the earth's land surface.

So, was it just luck, or is there something special about Antarctica?

With all these meteorites being collected in Antarctica, you might think that there is some sort of weird phenomenon going on, perhaps fireballs are preferentially focussed towards the poles. In fact, the very opposite may be the case. A recent study has suggested that there is a significant decrease in the number of fireballs at the poles compared to the equator [11]. This would suggest that meteorites are less likely to fall in Antarctica (and the Arctic) than anywhere else on the planet. So, how do we explain the amazing recovery statistics? Well, the simplest reason is that against a white background, a black rock is easier to spot. Also, one of the biggest issues with meteorites is the fact that they degrade fast in our oxygen-rich, terrestrial atmosphere. Meteorites often contain a lot of iron-rich metal and, to put it bluntly, they "rust" in the more hot and humid regions of the Earth. But Antarctica is a fridge, and while alteration does take place in the Antarctic climate, it is much slower than in many other places.

But cold, slow weathering is still not the whole answer. There is also a specific meteorite concentration mechanism in operation (Figure 12.4) [12]. In the area where ANSMET undertakes its collection activities, the enormous East Antarctic ice sheet, which flows outwards from the centre of the continent, is impeded by the Trans Antarctic Mountains. In these areas, deep ice may be forced to the surface. It will be carrying meteorites that fell elsewhere on the ice sheet and became trapped in the ice. At the same time, strong Katabatic winds from the continental interior

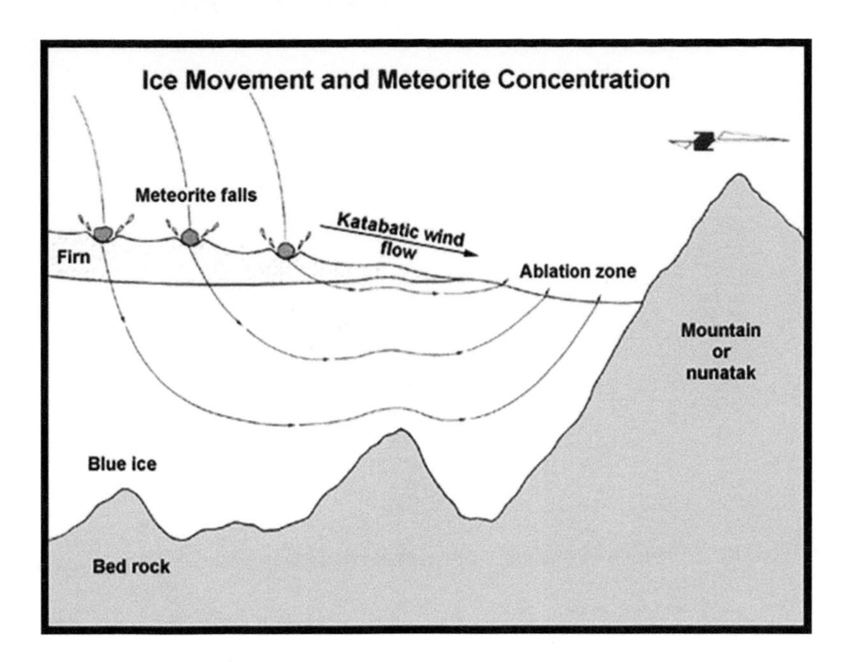

FIGURE 12.4 How flowing glaciers in Antarctica trap and concentrate meteorites. (Image: ANSMET/NASA.)

preferentially erode the ice and leave a deposit rich in meteorites on the surface (Figure 12.4). Areas where ice is being actively eroded are recognised on the surface as "blue ice" fields.

Unfortunately, the elegant scenario depicted in Figure 12.4 is not the whole story. It's more complicated than that. Isn't it always! A large number of Antarctic meteorites are found in areas with a high concentration of normal, terrestrial rocks (Figure 12.5). These collections of loose rocks are called moraines. Their formation has certainly involved some element of transport by moving ice, but at least on a local scale, wind transport may also be important. To spot the meteorites in these moraines normally requires a foot search, whereas blue ice recoveries are typically undertaken using Ski-Doos.

Antarctica is thus an amazing place to hunt for meteorites. There is of course just one problem. To benefit from this extraterrestrial treasure trove, you have to get yourself, your team, and all your supplies down there and out into the search area. From a logistical point of view, it's tricky, very tricky. Basically, only national research organisations can support the effort required. Unlike the intense private collection activities going on in the world's hot desert regions, in Antarctica, there are no commercial meteorite hunters at work. This is also in line with the Antarctic Treaty signed in 1959, which means that Antarctica should only be used for peaceful scientific purposes.

And it can be dangerous work too. In 1989, during a collection expedition to the Sør Rondane Mountains, located about 600 km from the original Japanese collection area in the Yamato Mountains, the Japanese team suffered a near fatal accident.

FIGURE 12.5 Unfortunately, meteorites in Antarctica are not just found all on their own, as is the case in Figure 12.1. They are often mixed in with lots of normal terrestrial rocks as part of a moraine. In this photo, taken during the 2014/2015 ANSMET expedition, the vast majority of the stones and boulders are not space rocks. It needs skill and training to spot meteorites amongst all the terrestrial rocks. (Photo: ANSMET/Vinciane Debaille.)

Following the end of a successful meteorite hunting season, the Japanese team strayed into a hidden crevasse field. One of their large, tracked snow vehicles (Figure 12.6) fell 30 m into a crevasse, and while trying to rescue his colleagues, another member of the team fell into an adjacent crevasse. The injured scientists were rescued a week later by helicopter and required hospitalisation back in Japan [13].

Undertaking a meteorite search in Antarctica is a complex operation [14]. An initial step is to identify promising field areas. This is done by locating blue ice fields in the vicinity of high relief areas that suggests meteorite concentration processes might be at work, as discussed earlier. An initial field survey is then undertaken to see whether the area is promising and makes sense from a logistical perspective. If all looks good, then a full field team (six to eight individuals typically) will go there to search for meteorites. Snowmobiles are an essential piece of equipment when searching for meteorites in Antarctica (Figure 12.7). Typically, most international expeditions will make use of them to cross an ice field when searching for space rocks, with vehicles generally spaced at about 30 m apart (Figure 12.8). When an area with a particularly high concentration is located, snowmobile traverses are replaced by searches conducted on foot.

When a meteorite sample is located, its GPS coordinates are recorded, it is photographed, field notes are made, and an identification number is assigned. The meteorite is then picked up using Teflon-coated tongs, placed in a pre-cleaned Teflon sample bag, and stored with the other samples collected in the field (Figure 12.9).

FIGURE 12.6 One of the large tracked snow vehicles used in some of the early Japanese Antarctic Research Expeditions. (Photo: the author.)

So, what happens next?

At the end of the field season, all the meteorites are shipped out of Antarctica for further study. In the case of ANSMET, that means they are sent to the Johnson Space Center in Houston, Texas. It is a long journey. From the field, the samples are returned to the McMurdo base in Antarctica, a cargo ship transports them to the Port Hueneme naval base in California, and finally a truck takes them on the final leg to Houston. And all the way along the route, they are kept frozen. Once at the Johnson space center, they are gently thawed and initial curation is carried out. The next step is to classify all these samples. It is a huge task which is jointly undertaken by scientists at the National Museum of Natural History (NMNH), part of the Smithsonian Institution in Washington, DC [15,16] and NASA scientists based at the Johnson Space Center. The results of these studies are then submitted, reviewed, and approved by the Nomenclature Committee of the Meteoritical Society (NomCom). Like any meteorite sample, these classifications will be entered on the Meteoritical Bulletin database. In addition, the ANSMET samples are published twice a year in the Antarctic Meteorite Newsletter [17]. Most other groups studying meteorites in Antarctica will operate a similar system.

Based on the information contained in these newsletters and on the Met Bull Database, scientists around the globe can then make official requests for sub-samples of the meteorites. In the case of the US collection, these requests are reviewed twice a year by an independent committee of scientists known as the Antarctic Meteorite

FIGURE 12.7 Snowmobiles are an indispensable part of any meteorite expedition in Antarctica. Here, a Ski-Doo is being used to pull a metal detection array as part of the UK initiative to search for iron meteorites that might lie below the ice surface. (Photo: Katherine Joy/Lost Meteorites of Antarctica.)

FIGURE 12.8 A Ski-Doo meteorite search conducted during the 2014/2015 ANSMET expedition (Photo: ANSMET/Vinciane Debaille.)

FIGURE 12.9 Meteorites being packed away in the field for the return journey to the Johnson Space Center in Houston. (Photo: NASA/ANSMET/Vinciane Debaille.)

Allocation Panel (AMAP) [18]. Here, I must declare a personal interest. I was a member of this allocation committee for three years from 2014. Back then, it was known as the Meteorite Working Group (MWG). It sounds like a pretty technical, and not very glamorous, activity and sometimes, like most jobs, that was true. But it was also a privilege to be involved in such an important scientific endeavour. The MWG went out of its way to make sure that scientists requesting material got what they needed to undertake their studies, while at the same time bearing in mind that material also needs to be preserved for future generations of scientists. It was a very positive experience and it goes without saying that I learnt an awful lot about meteorites while doing the job.

Japan and the US are not the only nations that have been actively involved in the collection of meteorites in Antarctica. Other countries that have searched for meteorites in Antarctica include: Belgium [19], China [20], Italy [21], South Korea [22], a European consortium known as EUROMET [23] in the early 1990s (Figure 12.10), and most recently, the UK [24]. You might think that having already collected 44,000 specimens that would be enough (and in fact there are a lot more in the pipeline that haven't yet been classified). But it's a bit like panning for gold. You have to trawl through a lot of specimens to find the rare and important samples. And there have been some pretty amazing meteorites located in Antarctica.

The first meteorite to be recognised as having a lunar origin, Allan Hills A81005 (Figure 12.11), was collected in Antarctica by ANSMET [25]. We now have a further 43 lunar Antarctic meteorites (Meteoritical Bulletin database February 2024).

FIGURE 12.10 Collecting meteorites in Antarctica is an international activity. Professor Rainer Wieler of ETH Zürich is seen here bagging a meteorite as part of the European Union funded EUROMET consortium in the early 1990s. (Photo: EUROMET/Ian Franchi.)

The paired Antarctic meteorites GRA 06128 and GRA 06129 are unique samples from an unknown asteroid that produced granitic crust similar in composition to that found on the Earth [26]. Asuka 881394 is another unique sample collected by the Japanese. It was originally thought to be a member of the HED suite of meteorites and so by implication originally from asteroid (4) Vesta. But further study, in particular oxygen isotope analysis, has revealed it to be a sample from a unique and as yet unidentified asteroid [27]. Then, as of February 2024, there have been 30 Martian meteorites collected in Antarctica. The most famous of these is Allan Hills 84001 (ALH 84001), which contains features that were interpreted as being evidence in favour of the existence of life on Mars in the distant past [28–31]. These extraordinary claims made a huge media impact back in 1996. It is fair to say that the evidence cited in favour of past Martian life found in ALH 84001 is now generally viewed with a great deal of scepticism by scientists. The search for life on Mars is

FIGURE 12.11 Allan Hills A81005 is a 31.4g Lunar Anorthosite Breccia found by ANSMET in the 1981–1982 field season. Randy Korotev reports: "*Allan Hills A81005 is the first meteorite to be recognized as originating from the Moon. It was suspected to be lunar in the field and when first examined in the lab (above). After many studies it was proven to be of lunar origin in the spring of 1983, a little over 10 years after the last Apollo mission to the Moon*" [25] (Photo: NASA/ANSMET.)

currently focussed on the results obtained by Martian orbiters and rovers. However, the widespread interest in Martian life that ALH 84001 created certainly helped NASA enormously in its quest to secure additional funding for Mars research. We will look at Martian meteorites in further detail in Chapter 15.

Compared to meteorites collected in hot desert regions, those from Antarctica tend to be much better preserved. But that begs the question, if meteorites from hot deserts are so weathered, why bother with them? Part of the answer is money. Commercial dealers can't get samples from Antarctica. But it is not just about financial factors. And course, not all desert meteorites are heavily weathered. Let's just say it's complicated! In fact, the world's hot desert regions are an increasingly important source of rare and scientifically valuable specimens. It's time to head back to the desert and take a closer look at what's going on.

13 Desert Gold

A vast sand-filled desert, mountainous dunes, and in the distance a slowly moving line of camels. The world's hot and inhospitable places have a romantic, if not mystic aura. Maybe you'd be surprised to know that they are also crammed full of amazing meteorites. A slight exaggeration perhaps, but it's time to explore the extraterrestrial possibilities of the world's desert regions. On the way, we will meet a legendary French flying author, join the most unsuccessful meteorite expedition of all time, and examine why some meteorites are worth more than their weight in gold.

No one understood the romantic, almost spiritual, dimension to space rocks better than the celebrated French author and pioneer aviator, Antoine de Saint-Exupéry [1]. His much-loved children's book Le Petit Prince is beautifully illustrated with numerous drawings of asteroids, all produced by the author himself (Figure 13.1) [2]. The scenario of The Little Prince involves an aviator whose plane breaks down in the desert, far from civilisation, and who then encounters the mythical Little Prince. The famous first words uttered by The Prince are "Draw me a sheep". The aviator has no problem fulfilling this request and the two then set off on a series of adventures that take place mainly on a succession of asteroids, each with a single, slightly bizarre inhabitant [3]. The Little Prince is a celebrated and classic children's story and helped to make Antoine de Saint-Exupéry a national hero in France. Sadly, he never lived to enjoy the success of the book, which was first published in April 1943. He was killed in an aircraft accident in July 1944, while on a reconnaissance flight from Corsica to collect intelligence information about German troop movements in southern France. The location of his lost aircraft and the circumstances of his death were matters of deep emotional interest in France for over 50 years, until the wreckage of his plane was finally located by a diver in May 2000. However, this has still not settled the speculation about the circumstances of the crash [4].

The story of The Little Prince was inspired by Saint-Exupéry's experiences as a pioneer aviator in Africa. In his book, Wind, Sand and Stars [5], he describes his adventures in the late 1920s as a student pilot with the Latécoère company, a predecessor of Air France, flying between Toulouse and Daker, in present-day Senegal. On one occasion, "*a minor accident*" forced him to land on a table plateau in the Rio de Oro region of Western Sahara. The local rock was a shelly limestone, very

DOI: 10.1201/9781003174868-13

FIGURE 13.1 The French aviation pioneer and author Antoine de Saint-Exupéry crash landed on a desert plateau in Western Sahara. There he found rocks he took to be meteorites. This encounter may have been part of the inspiration behind the children's classic Le Petit Prince.

flat and remote. Saint-Exupéry was convinced he was the first human to have visited the site. And then he noticed amongst the light-coloured rocks "*a hard black stone, the size of a man's fist*". He wondered what this black rock was doing there and had the following moment of inspiration. He thought of the flat tabletop mountain as if it were a sheet "*A sheet spread beneath an apple-tree can receive only apples; a sheet spread beneath the stars can receive only star-dust. Never had a stone fallen from the skies made known its origin so unmistakably*". This is a very beautiful and poetic idea. A natural tablecloth spread out to catch falling meteorites. But is it realistic? My own experiences in hunting meteorites in Morocco perhaps provide an insight into whether we should take Saint-Exupéry's suggestion at face value.

In the early 1990s, I went to Morocco with Open University colleagues Luigi Folco and Ian Franchi to hunt for meteorites (Figure 13.2). It was all the rage at the time. We were going as part of EUROMET, a European initiative, to search for space rocks in the world's hot and cold desert regions [6]. It was the brainchild of the late, great Colin Pillinger [7]. We started the journey in Milton Keynes, driving a very uncomfortable Land Rover down to Southampton to catch the ferry across the channel to Caen. Then, it was an all-day trek to Pau in the foothills of the Pyrenees to meet up with the eminent French planetary scientist, Michel Maurette. After an overnight stop, we drove on through Spain to the port of Algeciras. We took a ferry across the straits of Gibraltar, and finally, once we had arrived in Morocco, drove on to Rabat. We must have stopped somewhere on the way, but I have long since forgotten where it was. In Rabat, we sorted out the paperwork and met up with our

FIGURE 13.2 Getting your laundry dry in the desert while hunting for meteorites. The 1992 EUROMET expedition to Morocco. (Photo: Luigi Folco.)

official Moroccan guides. We then headed south through the high Atlas – amazingly beautiful scenery – and stayed overnight in the town of Erfoud. And finally, it was on to our search area, which consisted of flat, desert plains as far as possible from the Atlas Mountains as we could get.

At the start of the expedition, we were feeling pretty optimistic. The omens were good. For several years, there had been a steady stream of meteorites coming out of a variety of areas in Northwest Africa, particularly Algeria. Once we started meteorite hunting in earnest, we adopted the classic strategy of forming a line with each person separated from the next by 3 or 4m. We would then walk slowly in one direction looking out for any dark rocks that stood out against the light-coloured background of the desert floor. We spotted a lot of suspect rocks. But on closer inspection, all but one turned out to be dark, silica-rich terrestrial stones. We hunted for about two weeks, staying for a couple of days in each search area before moving somewhere new. Each area looked pretty much like the last, and all were essentially free of meteorites, or so it seemed to us. We soon began to doubt our skills as meteorite hunters. At the end of the two weeks, we headed back to Rabat in a pretty sombre mood. All we had to show for our toil was just one tiny sample. In Rabat, we met up with an elderly French geologist who had been working with the Moroccan geological survey for many years. We explained what we had been doing and she had a gentle laugh at our expense. It turned out that she had often been to the same area hunting for space rocks and had also found nothing. We started to feel a bit better about our failure. It seemed likely that flat areas close to the Atlas were just not the right place to be hunting for meteorites. Well, that has always been my excuse, anyway. And our one specimen? It was given the works back in Milton Keynes. It turned out to be terrestrial too. Never mind! (Figure 13.3).

FIGURE 13.3 In Morocco, we adopted the classic search strategy of forming a line and then slowly walking in unison (almost) looking for dark stones that stood out against the background of the lighter-coloured soil and rocks. It didn't work, we found absolutely nothing. (Photo: Luigi Folco.)

If Morocco had been the only string to the EUROMET bow, it would have been a very disappointing outcome. Happily, that wasn't the case. In association with the Italian Programma Nazionale di Ricerche in Antartide (PNRA), several hundred samples were collected in the Frontier Mountain ice field, Victoria Land, Antarctica, and four successful expeditions were run in the early 1990s to the Nullarbor region of Western Australia, in association with the Western Australian Museum, again yielding several hundred samples.

While I had initially been convinced that Saint-Exupéry had indeed found meteorites on the Rio de Oro plateau tops, as described in Wind, Sand, and Stars, doubts began to set in after our Moroccan failure. Black rocks on a light-coloured surface might be meteorites, but also might not. Meteorites are rare, and even in the most apparently favourable sites, you need a bit of luck and good fortune to find them. Nevertheless, it would be nice to know the truth about Saint-Exupéry's claim. Or perhaps not, it is a very poetic and inspiring idea.

Although many of the samples now coming out of the world's hot desert regions are collected by local people and often traded through dealers, organised scientific searches of such regions are still important. One of the most successful is the Omani-Swiss Meteorite Search Project and follow-up expeditions in Saudi Arabia, led by Beda Hofmann and Edwin Gnos from the Natural History Museums of Bern and Geneva (Figures 13.4–13.6). The project and its follow-up activities have been going since 2001, and the number of specimens collected so far is 7,282 from Oman and 82 from Saudi Arabia. As we have seen, meteorites generally fragment in the atmosphere before they impact the Earth's surface, and consequently, the number of actual meteorite events will be significantly less than the number of fragments

FIGURE 13.4 Hang on! Here is exactly what Saint-Exupéry was talking about. It's a black rock lying on a bed of light-coloured limestone and sandstone, and it is a meteorite collected by the very successful Omani-Swiss Meteorite Search Project. Looks like we should take the great man's claims a bit more seriously than I did a little earlier in this chapter. (Photo: Beda Hofmann.)

recovered from a collection area. The Omani-Swiss team estimate that the samples they have collected represent about 1,400 fall events. A classic approach to meteorite searching is generally adopted by members of the project, along similar lines to those we followed in Morocco (Figure 13.3).

Remember I mentioned that we had stopped over in the town of Erfoud during our travels south in Morocco. At the time of our visit, it was well known as an important centre for the sale of minerals and fossils. Some meteorites were also coming out of Erfoud, but it was nothing compared to the industry that has developed in the years since our, not so successful, expedition. Today, Erfoud is a major centre in the international trade in meteorites [8]. It is very big business, as we shall see.

People are often astonished when they find out that there is a lively trade in meteorites. But it's not so strange when you think about it. Like any precious and rare natural object, meteorites are collected by enthusiasts and sought after by museums and scientific institutions. In short, we are into the normal commercial environment of supply and demand. And since the demand is high and the supply is extremely limited, prices can be astronomical. Which is more than appropriate when you think about! Let's take lunar meteorites as an example. Yes, we have meteorites from the Moon! You would have thought that the 382 kg of Moon rock brought back by the Apollo astronauts would have provided more than enough material for everyone who wanted a piece of the action [9]. And in a certain sense, that's correct. If a scientist

FIGURE 13.5 Omani-Swiss expedition at work. Once a potential meteorite sample has been located, it needs to be carefully documented prior to being bagged up for shipment back to the lab. (Photo: Beda Hofmann.)

FIGURE 13.6 But classic searching on foot is not always the most effective way of dealing with the huge amount of terrain that needs to be covered. Searching from vehicles is also required, at least at the reconnaissance stage. (Omani-Swiss Project; Photo: Beda Hofmann.)

wants to study a lunar rock, in the first instance, they need to write to the Apollo Sample Curator at the Johnson Space Center, and set out in some detail what measurements they would like to undertake, and on what samples [10]. The process is fully open and transparent, and a sensible proposal will almost certainly be rewarded with a sub-sample of the requested material. But if you just want to purchase a lump of Apollo Moon rock to put on your mantelpiece for show, I am afraid that is just not going to happen. NASA guards its Moon rocks like the precious jewels they are. They are for science, not for display. They cannot be bought under any circumstances. There is no trade in Apollo samples because there is no supply. None whatsoever.

Unfortunately, that has not stopped some people trying to get hold of Apollo samples by illegal means. A notorious incident in July 2002 involved three university students who were working as summer employees (interns) at NASA's Johnson Space Center in Houston, Texas. Taking advantage of the access they had been given as part of their summer studentships, they stole a safe containing Apollo Moon rocks and lunar meteorites. They then tried to sell the material to a Belgium mineral dealer, who quickly alerted the FBI. As a consequence, they were arrested by agents posing as mineral collectors. The ringleader was given a custodial sentence of eight years and four months [11].

Initially, following the Apollo Moon landings, there was no legitimate way to own a piece of the Moon. However, while no one realised it at the time, this situation changed dramatically in 1979, and in Antarctica of all places. A Japanese team of scientists were hunting for meteorites in the Yamato Mountains and came across a 52 g rock of uncertain origin. In the usual way, it was given the technical designation, Yamato 791197 or Y-791197 for short [12]. Fast forward a few years, and a US team were out collecting in the frozen wastes of the Allan Hills, Antarctica. It's a place where many thousands of meteorites have been found, as we saw in Chapter 12. The team came across a strange looking rock containing centimetre-sized white crystals. It wasn't so big, weighing in at just 31 g. It would later be designated ALHA81005.

Along with all the other samples collected in the freezer that year, it was shipped back to Houston for further study. In due course, ALHA81005 was "classified". All that really means is that a sample is analysed using a range of techniques in order to decide firstly if it is a meteorite and if so what sort?

And that's when the fun really started, because ALHA81005 didn't look like any meteorite that had been seen previously. But what it did look like was the rocks that the Apollo astronauts had brought back from the Moon. A meteorite from the Moon? [13]. But that's impossible! The only way you can do that is if another big piece of asteroid smashed into the Moon and as a result of the impact, fired bits of the Moon into space. They would then zip around for a bit and finally land on Earth. In this case, in Antarctica. But as farfetched as it might seem, ALHA81005, after lots and lots of testing, and then even more testing, turned out to be the first verified lunar meteorite [14]. And once the cat was out of the bag, it was realised that the rock found back in 1979 by the Japanese in Antarctica, Yamato 791197, was also a lunar meteorite [15].

And as incredible as it might seem, pieces of the Moon have now been found in numerous places on Earth. At the time of writing (February 2024), there are 659 official lunar meteorites, with a total weight of over 1 metric ton [16]. Some of these were found in Antarctica by scientific expeditions, so are not available for sale. But a much larger proportion has been collected in hot desert regions, particularly the Middle East and North Africa, by local people and they are very much on the open market. In fact, slightly less than 6 kg of lunar meteorites have been collected in Antarctica. With prices ranging from $1,000 per gram, to well over $2,000 per gram, that means that even at the lower end of this scale, the sale of lunar meteorites has a total value of over 1 billion dollars. It is true that with the increasing mass of lunar meteorites now being found, prices are coming down, quite a lot in fact. But you get the general idea. The market for "lunars" represents big money by any standards. And that's just lunar meteorites. There are currently (February 2024) 376 officially recognised Martian meteorites [16]. And they go for very high prices too! No wonder there is now a thriving market in the sale of meteorites.

So how does this market work? Who is buying and selling space rocks? Interesting questions! You would be extremely surprised by some of the answers. Let's look first at where you can buy them. The most obvious sellers are the great auction houses. Meteorites are now sold on a regular basis by the likes of Sotheby's, Christie's, Bonhams, and Phillips [17–21]. They produce well-researched catalogues providing detailed information about the individual specimens in the sale. Not surprisingly, prices tend to be on the high side. Then, there's eBay and the like [22]. At any one time, there are large quantities of meteorites available for sale online. Many meteorite dealers do the majority of their trading this way. But how do you know that the material being advertised is legitimate? The short answer is, you don't. It is entirely possible that what you are purchasing is fake. But it is also unlikely. The vast majority of dealers are legitimate, and the market is self-policing. A rogue dealer will be spotted quickly, and once identified called out by other genuine traders who have a vested interest, and a pride, in keeping the meteorite market honest.

The annual mineral and meteorite shows, such as Tucson [23], or the more specialist Ensisheim meteorite fair (Figure 13.7) [24], are also great places to buy meteorites and chat with dealers. The Tucson Gem and Mineral Show [23] takes place annually in late January/early February. It is undoubtedly the world's most important venue for trading extraterrestrial rocks. By all accounts, it's a pretty amazing event, with dealers and collectors coming from all over the world. The range and quality of the material on show also attracts interest from major international museums. Tucson and Ensisheim are important annual events that bring dealers and collectors together.

However, it is also important to put this trade into a scientific context. The meteorites being bought and sold in Tucson and elsewhere are from non-Antarctic regions, particularly the desert areas of North Africa. As important as these samples undoubtedly are, they are still less significant in terms of numbers, and are often more weathered, than those collected in Antarctica by organised scientific expeditions.

But looking at specimen numbers alone is misleading. Back to those lunar and Martian samples again. Of the 659 officially recognised lunar samples (all are finds) on the Meteoritical Bulletin Database (February 2024), 615 specimens, or 93% of the world's collection, are non-Antarctic finds. And it's the same for Martian meteorites. Out of a global total of 371 finds (we will leave out the five non-Antarctic falls for now), 341 specimens were recovered from non-Antarctic regions. That means that only 8% of the world's collection of Martian rocks come from Antarctica. And that is pretty strange, because as we saw Chapter 12, Antarctica has provided 60% of all the world's meteorite samples. This begs the important question: why are so many lunar and Martian meteorites being found in non-Antarctic areas and so few (it would appear) in Antarctica?

Often when trying to make sense of the meteorite find numbers, it helps to also look at the statistics from meteorite falls. These should be less biased and more

FIGURE 13.7 Wood carving depicting the fall of the Ensisheim meteorite on 7 November 1492. The meteorite is now on show at the local museum and an international meteorite fair is held in the town each year. (Image: H. Raab/Wikipedia.)

representative of what is actually coming through the atmosphere, rather than how good collectors are at spotting space rocks. But unfortunately, this doesn't help much. The truth is, there are no lunar falls [25]. None! No one really seems to know why. There are five Martian meteorite falls and yet the total number of Martian meteorites (falls and finds) is only 376 compared to 659 lunar specimens (February 2024). Based on the ratio of Martian falls to finds, there should be about nine lunar falls. Strange! Perhaps there have been some recent lunar falls and they have gone unrecognised. It's possible, but not likely. It might be worth curators having a rummage through their collections and taking a look at some of those "meteorwrongs". There might be a lunar fall in there somewhere. But an even more interesting speculation is that the statistics suggest a lunar fall is now overdue. When that happens, it will be big news indeed!

But back to those weird statistics. It seems possible that the only reason so many lunar and Martian rocks are being found in hot desert regions is down to commercial interests. Ordinary chondrites have a low resale value. This probably means they are being left behind, whereas lunar and Martian samples, which fetch much higher prices, are being collected and sent to market in significantly higher numbers. It is the same with achondrites in general. These are highly valuable commodities and are being recovered at a prodigious rate as a result of their commercial value. This is generally good news for science as we are getting access to some very important space rocks. But it does mean that the collection statistics for hot desert finds need to be viewed with a great deal of caution.

An important means by which meteorites are bought and sold is through private one-to-one deals. As in all markets for valuable objects, there are wealthy collectors who are willing to pay high prices for unique specimens. The dealers know who they are and the sales go unrecorded. Is this the biggest part of the meteorite business? Perhaps, but no one really knows.

I would be giving a false impression not to mention that scientists and museum curators also regularly buy specimens. And here's why. Let's say you want to conduct a detailed investigation into the origin of a key group of meteorites. You will make written requests to a range of scientific providers, explaining what you want to do and how you will do it. It may take a few months, but eventually you will receive your samples at no cost to your institution. This is science after all! Governments and ultimately taxpayers funded the expeditions that collected this material, so it is not surprising that scientists get it at no extra cost. But key specimens will almost certainly not be available through this route and will need to be purchased from dealers. Outrageous! you might think. Science is being ripped off by commercial interests. And there are some scientists who have sympathy with that view. But most don't. It's all down to a sort of symbiosis. Because meteorites are so rare and precious, there is a market for them. That fuels a much higher level of searching in hot desert regions than could ever be undertaken by government-funded expeditions. The result is that there is a greatly expanded range of meteorite types available for study than would otherwise be the case. There are a number of very important samples that have come to light by this route. In Chapter 15 we will look at a key Martian meteorite that has been nicknamed "Black Beauty", its official title is NWA 7034 [26]. It is a unique sample with important implications for our understanding of Martian geological evolution. But it was not found by an official scientific search expedition. It was collected by local people in North West Africa and traded through the commercial dealership network.

At this stage, you might be asking how do you become a meteorite dealer? Are any qualifications needed? Is there a training programme? Basically, anyone can be a dealer, but you do need some prior knowledge of, and interest in, extraterrestrial rocks. All sorts of people are involved in the meteorite trade, from Indiana Jones types, to people who could equally be collecting stamps. Some dealers are highly commercially focussed, while others are really not in it for the money, but are just trying to cover expenses. Most have a positive attitude to the science, and on a number of occasions, I have been given key samples free of charge. It can also be a dangerous trade. Meteorites don't just land in politically safe countries. Some dealers have fallen foul of the local authorities and ended up in prison [27].

The relationship between dealers and scientist is not always completely harmonious, of course. But most researchers recognise that the commercial trade in meteorites has resulted in a much greater availability of rare space rocks than would otherwise be possible. And there are also logistical realities. A determined meteorite dealer can often get hold of a specimen that has fallen in a remote area much faster than would be possible through more conventional channels.

Finally, it's worth remembering that while meteorites recovered from hot desert regions can sometimes be very weathered, there is in fact no such thing as a totally pristine meteorite. Even the freshest fall, like Winchcombe, has been contaminated by interaction with the atmosphere, as it plunged to Earth on arrival. If you want totally pristine samples, then you need to travel into space to collect them. It is not as farfetched as it sounds. We are now going to examine the increasingly important field of extraterrestrial sample return.

14 Capturing Meteorites in Space

Waiting for meteorites to arrive on Earth can be a bit of a frustrating business. Long periods of boredom, short periods of terror, as we have seen. Why not head off into space and pick up some samples instead? It's an increasingly popular way of doing things. Mind you, it is also a bit pricey! But there are some very big advantages, as we shall see.

There can be a sort of menacing feeling when you talk to people about asteroids. The pertinent question seems to be: "Will they get us the way they got the dinosaurs?" You know the sort of thing. And of course, it is natural to worry about such existential threats. As a community perhaps, planetary scientists are guilty of playing up this very remote possibility. It captures people's imaginations. As depicted in blockbuster movies like Deep Impact [1], or Armageddon [2], asteroids are the real villains, always on a crash course to destroy planet Earth. It needs a bunch of renegade heroes to save humanity from certain destruction. You've seen the movie, got the T-shirt! But as we have seen, there is a lot more to asteroids than just their Hollywood persona as potential planet pulverisers. Unfortunately, it is so easy to get typecast these days!

As we discussed earlier in this book, some asteroids very likely formed in the icy outer reaches of the Solar System. These are the carbonaceous chondrites and they may have been the carriers of water and other volatiles to the early Earth, and the other inner planets. This is certainly a strong possibility [3]. But perhaps we could be accused of "bigging-up" this aspect of asteroids a little too much. Some scientists argue that there was always enough water hanging around in the inner Solar System to more than account for Earth's oceans [4]. And so there remains no certainty that primitive asteroids were needed to get life going on Earth. It's all very confusing.

If the threat from killer asteroids has been overstated and they didn't carry water and organics to Earth, what use are they? Interesting curiosities at best, ugly lumps of rock at worst? Having waded through the previous chapters of this book, I am sure you don't think that. Of course, asteroids are the real deal. They are truly fascinating and amazing objects, no ifs, no buts. Here's a little recap of why they are so important.

DOI: 10.1201/9781003174868-14

Unlike the planets, asteroids are frozen in time. Planets like Earth change and rejuvenate. They evolve and renew themselves. Erosion by wind and water, deposition by rivers, eruption of volcanoes, the slow building of mountain chains. These processes mean that planets are constantly evolving. But asteroids are witnesses to the Solar System's distant past. They formed within a few million years of Solar System formation, and have changed very little since. As we have seen, a unique set of objects within some primitive asteroids, known as calcium-aluminium-rich inclusions, CAIs for short, record the oldest Solar System ages. When scientists tell you that our Solar System is 4,567 million years old, that age came from dating CAIs in meteorites. Despite the uncertainty concerning meteorites as bringers of water, they remain scientifically important and fascinating objects. And in fact, they probably did bring the water anyway!

But you could argue that while they might be important, we do have rather a lot of samples from asteroids in the form of meteorites. And in some ways it's hard to disagree with that. However, as you might have guessed, this chapter is going to look at the case for flying space missions to asteroids, collecting samples directly from their surfaces, and then bringing them back to Earth (Figure 14.1). Now why would you want to do that if we already have so many bits of asteroids in the form of meteorites? It doesn't seem to make much sense. Only it does really.

I know it's a bit late in the book to admit it, but there are a few problems with meteorites. We do have a lot of them for sure, but they are filtered you see. Only the really strong ones can make it through the rigours of atmospheric entry. The big, crumbly, soft, fragile meteorites are often destroyed. And our idea of what meteorites are like

FIGURE 14.1 The sample return canister from the JAXA Hayabusa2 spacecraft was recovered from the Australian desert in December 2020. (Image: JAXA.)

is very dominated by observations made on the big ones that survive down to the surface. After all, they are the samples we meet in the world's museum collections.

But we now know that the bulk of extraterrestrial material arriving on Earth comes in the small size range of a few millimetres or less. The big chunky ones are less important when it comes to the overall mass of material arriving on Earth. Now, here's the thing. The small ones are very different in composition to the big ones. Carbonaceous, hydrated types, hydrated because some of their minerals contain water, and which are from further out in the asteroid belt, dominate the micrometeorite population. In comparison, the bigger meteorites are dominated by the ordinary chondrites that come from the inner part of the asteroid belt. In other words, meteorites are biased. Why is this? It has a lot to do with their strength, or lack of it. As they descend through the atmosphere, the forces on a meteorite start to build up. If they are weak and friable, they fall apart more easily than the tough, rocky ones. This demonstrates that meteorites, at least the big ones, are not representative of the materials from which many asteroids are composed.

And that's not all!

As they come hurtling through the atmosphere after being in the hard vacuum of space, meteorites act like sponges and absorb gases and moisture from the air. So, meteorites are already terrestrially contaminated before they even land on Earth. Now a freshly fallen meteorite is, well, very fresh, compared to a meteorite that has been lying around on the surface of a desert for 10,000 years, but it is not pristine. Its composition has already been modified by interaction with the atmosphere.

And while we are talking about problems here, there is another one we need to discuss. Naturally, one of our aims is to match meteorites to asteroids. But it's tricky, or can be. Due to a variety of processes, particularly resulting from bombardment by energetic particles streaming off the Sun, the outer surface of an asteroid can be changed and modified compared to its interior. This process is called space weathering [5]. This outer surface is what a spacecraft, or an Earth-based telescope, sees when it is making its observations of an asteroid. Now, what we want to do is match the asteroid to the meteorite. An excellent way to do this is to look at the light reflected from the fresh surface of a meteorite measured in the lab, and then compare it to observations of the light reflected from the surface of the asteroid itself. Normally, this is done by looking at wavelengths in the visible and infrared regions of the light spectrum [6]. But while you can knock the fusion crust off a meteorite, you can't normally do the same to an asteroid (this is not strictly speaking true, but we will get to that in a bit). This means that you will generally not be comparing like with like. It would be a lot better if you could collect a sample from an asteroid and return it to Earth for detailed studies. Expensive, but nice!

And that's exactly what we have started to do. In fact, we have been doing it for years. As we shall see. This whole process is known as "sample return". The very first set of sample return missions were of course the Apollo Moon landings. Between the launch of Apollo 11 on 16 July 1969 and the return of Apollo 17 on 19 December 1972, six missions (Apollo 11, 12, 14, 15, 16, and 17) brought back to Earth a total of 382 kg of Moon rock (Figure 14.2) [7]. It is fair to say that these samples have utterly changed our understanding of the early history of the Earth and

FIGURE 14.2 Apollo 11 Astronaut Buzz Aldrin on the Lunar Surface. The six Apollo missions brought back a total of 382 kg of Moon rock. From the point of view of returned mass, the Apollo missions were the most successful sample return activity so far undertaken. Of course, the sheer amount of sample recovered is not the only criteria for success, far from it. But then again, it is still a pretty impressive achievement.

Moon. It was analysis of lunar rocks that led directly to the realisation that the Moon formed as a result of a giant impact between the proto-Earth and another world, probably similar in size to Mars [8]. Detailed study of lunar rocks has helped us to understand the processes that shaped the early Solar System, in particular the early high rates of asteroid bombardment that would have affected the Earth, as well as the Moon [9]. The analysis of lunar soils returned by the Apollo astronauts has provided a vital resource in terms of understanding the flux of high energy particles that come from the Sun, often referred to as the Solar Wind [10].

The Apollo missions were not the only ones to return lunar samples to Earth. The Soviet space programme scored a notable success in 1970, when its fully robotic Luna 16 mission returned 101 g of lunar soil to Earth. This was followed by two further successful robotic missions, Luna 20 in 1974, which returned 55 g of soil, and Luna 24 which returned 170 g in 1976. In November 2020, 2 kg of lunar soil was returned by China's robotic Chang'e 5 mission [11].

Returning large amounts of extraterrestrial sample material is very nice, of course, but not always possible, or even necessary, as two important NASA missions demonstrated. The first of the two to be launched was Stardust in February 1999.

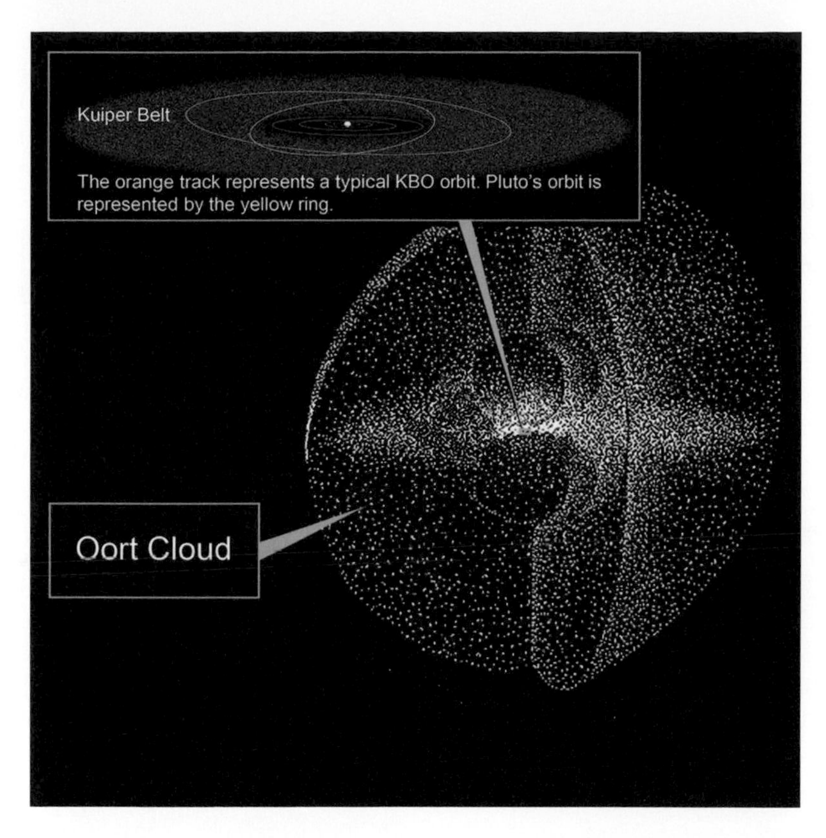

FIGURE 14.3 Diagram showing the relationship between the Kuiper belt and Oort cloud. (Image: NASA.)

Its main objective was to collect material ejected from Comet Wild 2 and return it to Earth. It wasn't going to be easy. Comets have been described as "cosmic snowballs", frozen balls of dust, gas, and ice, essentially unchanged since the birth of the Solar System [12]. They are thought to originate either in the Kuiper belt or the more distant Oort cloud. The Kuiper belt is a doughnut-shaped ring of icy bodies that starts roughly at the orbit of Neptune and extends far out into the outer reaches of the Solar System [13]. The best-known object in the Kuiper belt is Pluto, but it contains a great variety of other icy worlds, including comets. The Oort cloud is very different from the Kuiper belt. It is much, much further away from the Sun, and instead of being a doughnut-shaped ring, it has the form of a diffuse sphere that encloses the rest of the Solar System (Figure 14.3) [14].

As a comet moves inwards towards the Sun, it begins to warm up and produce a cloud of gas and dust, known as a coma. This happens because the ice it is made of sublimes and the dusty material found in the ice is released from the comet's frozen nucleus. It was these released dust grains that Stardust was designed to capture. But there was a problem. As they are ejected from the comet, these particles are moving at high speeds and would be completely destroyed on impact with the surfaces of the spacecraft, unless they are slowed down in some way. To avoid this problem, Stardust made use of a unique capture material called Aerogel, a translucent silicon-based material that is 99.8% air [15]. Aerogel is crazy stuff; when you hold it in your hand, you have the impression that it is almost not there. Particles entering the Aerogel capture cell are slowed down and brought gently to a halt. A distinctive track is created which can be used to locate the particle.

Stardust encountered Comet Wild 2 in 2004 and then two years later the sample capsule was successfully returned to Earth. And what did Stardust find? The headline discovery was the presence amongst the particles of fragments of chondrules and calcium-aluminium-rich inclusions (CAIs). As we saw earlier, CAIs may have formed close to the Sun. So, what were they doing in a comet that was coming from the outer reaches of the Solar System? For most scientists, this was a bit of a shock. It meant that material in the early Solar System was being transported huge distances, and not just inwards towards the growing Sun. There was clearly significant outward transport as well. And chondrule fragments, generally viewed as being characteristic features of meteorites, point towards the possibility that comets and asteroids have more in common than had previously been suspected. Stardust was a huge success, and yet the amount of material returned to Earth was only about one-thousandth of a gram (1 mg) [16] (Figure 14.4).

Launched in August 2001, the NASA Genesis mission retuned even less material than Stardust. Its objective was to collect particles ejected from the Sun, the so-called Solar Wind [17]. The concept behind Genesis was that it would travel to

FIGURE 14.4 NASA Stardust spacecraft making a close approach to Comet Wild 2 (Artist's concept). As depicted in this image, dust and particles are being actively ejected from the comet's coma, as the frozen comet nucleus is warmed by the Sun. (Image: NASA.)

the L1 Lagrange point, where the Earth and Sun's gravity balance [18]. The spacecraft could effectively hover there and open up an array of collectors that would then trap the energetic particles coming from the Sun's outer margins. These collectors would then be returned to Earth for intensive laboratory analysis. As the Sun represents 99.8% of the material in the Solar System, Genesis was a mission that had the capability of answering some very fundamental questions about our star and how it formed. Unfortunately, things didn't go quite according to plan.

On its return to Earth in 2004, an incorrectly installed sensor resulted in the main parachute failing to deploy and the sample capsule "augered in", hitting the Utah desert at an unintendedly high speed [19] (Figure 14.5). The capsule, with its precious cargo of solar wind–implanted surfaces, was supposed to have been captured, in midair, in a net, by helicopters flown by Hollywood stunt pilots [20]. I kid you not! Instead, the capsule split open and all its smashed contents were covered in a layer of fine desert dust. It was a catastrophic end to such an important mission. I watched the return live at the Open University and my two daughters were there too. As it became clear that something had gone badly wrong, my eldest said very loudly, "Daddy, was that supposed to happen?" I replied something along the lines of "um, probably not".

So, was that the end of the Genesis mission? Reduced to a heap of junk, half-buried in the soil of the Utah desert (Figure 14.5). Don't be daft, it's NASA! After, I would imagine, some well-chosen expletives, the scientists got on with the serious job of rescuing the mission. Sorting out the pieces of the broken collector plates required some very advanced jigsaw puzzle skills [21]. But finally, once the desert dust had

FIGURE 14.5 NASA Genesis mission sample return capsule after an unconventional landing. Believe it or not, NASA scientists succeeded in salvaging a significant part of the mission objectives, despite the crash. (Image: US Air Force.)

been slowly and painstakingly removed, some important measurements were made. A huge instrument at the University of California, Los Angeles (UCLA), called the Mega-SIMS, analysed the oxygen isotope composition of the implanted solar wind [22]. It was one of the most important measurements of the mission. When the data had been collected and suitable corrections applied, it turned out that the isotopic composition of the Sun was very different from that of the Earth. It was a result that had been predicted a few years earlier by the celebrated scientist Robert Clayton of the University of Chicago [23]. The fact that the Earth and Sun are so different once again emphasises the unique nature of our home planet.

Snatching success from the jaws of disaster was also the theme of the Japanese Space Agency (JAXA) Hayabusa mission to the near-Earth asteroid 25143 Itokawa (Figures 14.6). It was launched in 2003, touched down on the surface of the asteroid in November 2005, left the asteroid in 2007, but didn't get back to Earth until June 2010. The story of Hayabusa's journey was like something from Homer's Odyssey. Anything that could go wrong did go wrong, and yet just when it seemed to be game over, the engineers pulled a new trick out of the bag and saved the day [24]. On its way to Itokawa, the largest solar flare ever recorded damaged its solar cells. At the asteroid, some of Hayabusa's space hardware failed. Its attempt to launch a small probe onto the asteroid's surface failed. The touchdown procedure to sample the asteroids surface, well, it didn't go quite according to plan, and that's putting it mildly. A second touchdown was even less successful. The problems continued, more or less, all the way back to Earth. But the JAXA engineers just wouldn't give up [25]. The sample capsule parachuted successfully to Earth in June 2010 and was recovered from the Woomera landing site in Australia, and returned to the JAXA curation facility in Tokyo. It was by no means clear that Hayabusa had collected any material from Itokawa, but it had [26,27]. There were more than enough grains to demonstrate that the predicted meteorite match to Itokawa, made by remote

FIGURE 14.6 Artist's view of the JAXA Hayabusa spacecraft approaching asteroid Itokawa. As can clearly be seen from this image, Itokawa is a "rubble pile" asteroid, comprising an assemblage of loosely held-together angular boulders, stones, and dust. (Image: JAXA.)

sensing measurements, was indeed correct. Itokawa had shown a very good match to LL group ordinary chondrites and the particles returned by Hayabusa were? Yes, LL group ordinary chondrites [28].

In the end, Hayabusa was such a big success (they should make a movie about it) that a sequel was needed, and duly arrived. But the Hayabusa2 mission initially seemed to be reading from a very different script, less a thriller, and more just a boring sequel. You have seen it all before. An exciting debut film, lots of spills and thrills. But oh dear! the sequel, dull as dishwater and utterly predictable. I remember seeing the Hayabusa2 scientists at the 2019 Lunar and Planetary Science conference. The results were coming in from the remote sensing phase of the mission. It didn't look good. The target asteroid Ryugu seemed to be a carbonaceous chondrite type that had been heated up, potentially being less pristine than anticipated. But in the end, Hayabusa2 was not so much a dull sequel, but more an epic blockbuster. Welcome to probably the most successful asteroid sample return mission so far [29].

Launched in December 2014, the Hayabusa2 spacecraft reached asteroid 162173 Ryugu in June 2018 (Figure 14.7). For the next year and a half, it studied the asteroid in great detail, using a range of instruments. The data was relayed back to Earth and an assessment made of the likely composition of the asteroid. Asteroid Ryugu is a C class asteroid which means that it most closely resembles the dark carbonaceous chondrite meteorites, such as Murchison, Allende, and Winchcombe. They are crammed full of organic compounds, and some contain a lot of water. But the initial news from Ryugu was not good. Yes, it was a dark carbonaceous type, and it did contain water, a bit. But the spectra suggested that it was a heated type, not the pristine material everyone had been hoping for [30]. It was hugely disappointing. For reasons that are still not yet well understood, some carbonaceous asteroids were heated to relatively high temperatures of at least 500°C in the early Solar System, and partially

FIGURE 14.7 Asteroid 162173 Ryugu, the target of the JAXA Hayabusa2 mission, compared to asteroid 101955 Bennu, which was studied and sampled by the NASA OSIRIS-REx mission. Both asteroids have a similar shape and are rubble piles. They are both dark, water-bearing bodies, with compositions that are similar to carbonaceous chondrite meteorites. Ryugu is a little larger than Bennu. (Images: JAXA and NASA.)

to fully dehydrated [31]. Their organic material was also modified. The meteorite equivalent of these heated asteroids are known as CY chondrites [32]. Spending billions of Yen to bring back a piece of CY chondrite was not really the scenario JAXA would have hoped for, but they were making the best of it.

Unlike the first Hayabusa mission, operations at Ryugu had gone relatively smoothly [33]. It carried four small surface rovers, three of which worked well, the fourth was less successful. Hayabusa2 performed two sample retrieval touchdowns. The first picked up surface material. The second sampled material close to a crater that Hayabusa2 had created using a small projectile. The idea was to break through any space weathered surface and collect material more representative of the interior of the asteroid. That procedure went well too. Hayabusa2 left Ryugu in November 2019 and returned its sample canister to Earth on 5 December 2020. This was during the global COVID pandemic and so it was nice to get a bit of good news in what had been a pretty rubbish year. I remember watching the live feed from JAXA, the Japanese Space Agency, as the sample return capsule from the Hayabusa2 spacecraft blazed a trail in the skies above Australia. It landed safely in the Woomera Test Range and was located within a few hours by the recovery team (Figure 14.1).

The sample capsule was then taken to a "quick-look facility" close by and later transported to the JAXA campus in Sagamihara, Japan [34]. Once properly opened, a very intensive phase of sample characterisation studies began. Inside the capsule was 5.4 g of what the spacecraft operations at Ryugu suggested would be heated and partially dehydrated carbonaceous chondrite. Only it wasn't. JAXA had hit the jackpot. What was actually inside was gold dust, or the scientific equivalent of it. It turned out to be material similar to a rare meteorite type known as CI chondrites [35,36]. To understand why this was such a welcome surprise, we need to talk a bit about Ivuna-type carbonaceous chondrites. CIs for short.

CI chondrites have a composition that matches that of the Sun [37,38]. And as the Sun represents about 99% of the mass of the Solar System, it means that these asteroid fragments are basically the stuff from which the Solar System formed, apart from gases such as hydrogen and helium of course. When you hold such a meteorite, you basically have our Solar System in your hands. Unfortunately, this type of asteroid is very soft and crumbly, and only very rarely survives atmospheric entry to reach the Earth's surface as a meteorite. Plus, when it does arrive from space, it quickly interacts with Earth's atmosphere and that changes its composition. Imagine the shock, surprise, and joy, when, instead of returning a heated carbonaceous chondrite sample as expected, the sample cannister turned out to be full of this very rare and precious material. Ryugu samples, being essentially uncontaminated, are already helping to make sense of how meteorites become contaminated due to interaction with the Earth's environment, both during and after atmospheric entry [39].

And here is another important question raised by the return of CI-type material from asteroid Ryugu. Are water- and organic-rich asteroids like asteroid Ryugu more common than the meteorite record suggests? [40]. There may be a lot of CI-like material out there, it just doesn't make it down to the surface of the Earth as large lumps, because it is so weak and fragile. Hayabusa2 is likely to change many of our ideas about the importance of these old, water- and carbon-rich asteroids. And with this precious material safely in our laboratories, we have a much better chance of working out whether such asteroids really did bring the ingredients of life to Earth.

While Hayabusa2 was the first mission to successfully return material from a dark, primitive, organic- and water-rich asteroid, it is not the end of the story. Far from it. On 8 September 2016, 20 months or so after the launch of Hayabusa2, the NASA OSIRIS-REx spacecraft blasted off from Space Launch Complex 41 at Cape Canaveral, Florida. It was on its way to study and sample the dark, primitive asteroid Bennu [41]. Asteroid Ryugu, the target of the Hayabusa2 mission and Bennu show many similarities (Figure 4.7). While asteroid Ryugu is slightly larger than Bennu (896 m equatorial diameter compared to 565 m for Bennu), both are so-called rubble-pile asteroids and have a similar spinning top shape (Figure 4.7). Rubble pile asteroids [42] are basically a heap of smashed up rocks and finer particles, weakly held together by gravity. There are a lot of gaps and voids between the fragments, and so the overall density of these asteroids is very low. OSIRIS-REx arrived at Bennu on 3 December 2018, and then commenced a two-and-a-half-year program of detailed observations and measurements to determine its composition and structure [43].

On 20 October 2020, after a lot of detailed reconnaissance work, the OSIRIS-REx spacecraft finally headed down to the surface of Bennu to collect material from a small crater-like feature called Nightingale. This was not a rapid process; the descent took over four and a half hours. And it was also a potentially dangerous manoeuvre for the spacecraft. Bennu turned out to have a much more boulder-strewn and rough surface than had been anticipated prior to the specraft's arrival (Figure 14.7). One particularly large boulder close to the touchdown site was two storeys high and as a consequence, was nicknamed "Mount Doom" [44]. But in the end, it all went well. The video sequence taken at the time of touchdown is particularly spectacular, with a cloud of debris flying in all directions [45]. It clearly illustrates the loose and unconsolidated nature of the asteroid's surface.

The OSIRIS-REx spacecraft left Bennu and began its return journey to Earth on 10 May 2021 [43]. Reentry of the sample capsule took place on 24 September 2023, with a successful touchdown and recovery in the Utah desert (Figure 14.8) [46]. The capsule was then returned to the Johnson Space Center, where it was opened in a nitrogen-filled cabinet. This initial curation phase has been able to show that the spacecraft collected more than 60 g of material from the surface of Bennu. As I write this (February 2024), the material collected by OSIRIS-REx is being actively analysed in laboratories all round the world, including our own at the Open University. It is a very exciting time, and we are collecting new information about the mineralogy and composition of Bennu on a daily basis. Like the Hayabusa2 mission, OSIRIS-REx will provide important new insights into how the Earth became a habitable planet.

Despite all the problems and expense, sample return has already provided us with a wealth of new and exciting information about the early Solar System. And the good news is that there is a lot more to come. Mars remains a major exploration target

FIGURE 14.8 Members of the OSIRIS-REx recovery team examine the sample return capsule just after its safe landing in the Utah desert on 24 September 2023. Dante Lauretta, principal investigator of the mission, is on the right. (Credit: NASA/Keegan Barber.)

for most of the world's space agencies. Returning material from the red planet is not imminent but is certainly getting closer [47]. Our first samples from Mars may come not from the surface of the planet, but from Phobos, one of its two moons [48]. JAXA is planning to launch a sample return mission to Phobos in 2024 called MMX **M**artian **M**oons e**X**ploration [49], which will bring back sample material five years later. It is a neat mission for two reasons: firstly, because it will almost certainly recover material from the red planet, and secondly, it stands a good chance of working out how Phobos formed. Let's take a look at that second objective first (Figure 14.9).

Let's be honest about it, no one has much of a clue how Phobos formed. One theory says it's a captured primitive asteroid [50]. Well, it does look like one. Another theory claims it formed after a giant asteroid slammed into Mars [51]. It's very confusing. If Phobos really did form from a big impact event on Mars, then JAXA will be the first to bring back samples from the red planet. That's a big win. If it's a captured asteroid that will sort out the mystery, return some cool samples, including possibly bits of Mars. Another big win. A very smart mission concept. Why didn't someone think of this earlier? They did! Phobos-Grunt was a Russian mission with the objective of returning a sample from Phobos. It launched in 2011, and although it got into orbit around Earth, the rocket burn needed to send it on its way to Mars never happened, and eventually, the spacecraft had to be destroyed [52]. In the longer term, NASA and ESA are planning a joint series of missions to return the Martian samples already collected by the Perseverance Rover at Jezero crater [53].

FIGURE 14.9 Phobos, one of the two moons of Mars and the target destination of the JAXA MMX mission. No one really knows how Phobos formed. It could be a captured asteroid, or it could have formed following a large impact into the surface of Mars. The JAXA MMX mission will return samples from the surface of Phobos later in the 2020s. We could be in for a few surprises! (Image: NASA.)

The return date is currently set for 2033. But these are complex missions, and the timeline looks ambitious. However, there is no doubt that one day we will have samples collected and returned from Mars.

In the meantime we already have lots of lovely meteorites from Mars to keep us busy. But how do we know they are from Mars? That's all coming up in the next chapter.

15 The Chances of Anything Coming from Mars

In the celebrated science fiction novel The War of the Worlds (1898) by H.G. Wells, Martians invade Earth. In real life, thanks to NASA landers and rovers, we thought we were the first to make the journey between the two planets. Only we were wrong! The Martians arrived on Earth long before we got to Mars! And if that doesn't make sense, don't worry, it's all explained in this chapter.

Up until now, we have been looking mainly at meteorites that are essentially bits knocked off asteroids, relatively small objects that normally hang out in the asteroid belt between Mars and Jupiter. But that is only part of the story. In Chapter 13, we also looked at lunar meteorites. These clearly show that we have samples from some of the bigger bodies in the Solar System as well. We also briefly discussed Martian meteorites in Chapter 12. But how do we really know they are from Mars (Figure 15.1)? Let's take a closer look at the evidence.

Martian meteorites are sometimes referred to as SNCs. SNC is an acronym derived from Shergotty, Nakhla, and Chassigny, which are three of the five meteorites in this group that have been observed to "fall" to Earth. The other two falls are Zagami and Tissint. In a way, the group should now be more correctly called SNCZTs, or something like that. Not surprisingly that name hasn't caught on and most people just refer to them simply as Martian meteorites, which brings us back to the original question: how do we know they are from Mars (Figure 15.2)?

There are a number of lines of evidence linking Martian meteorites to Mars that were nicely summarised in a paper by Professor Hap McSween in 1994 [1]. Martian meteorites, with a few notable exceptions, generally show young crystallisation ages, ranging from about 2,300 to only 180 million years before present [2]. In comparison, meteorites from melted asteroids have very old ages, close to that of the Solar System itself, which formed 4,567 million years before present [3]. This evidence suggests that the body from which these meteorites originated must be large, i.e., a planet, in order to sustain long-term volcanic activity [1]. However, the meteorite ALH 84001, famous for the controversial claims that it contained evidence for

DOI: 10.1201/9781003174868-15

FIGURE 15.1 A Martian fighting machine in Woking, Surrey. A sculpture inspired by the famous science fiction novel by H.G. Wells. (Photo: Wikipedia.)

Martian life (Chapter 12), is an exception, with an age of approximately 4,100 million years before present. The important Martian meteorite NWA 7034 also has a relatively old age, but more on that in a bit [2] (Figure 15.3).

The two NASA Viking landers, which arrived on Mars in the summer of 1976, made measurements of the composition of the Martian atmosphere. Then, in 1983, analyses of trapped argon, krypton, and xenon in the meteorite EETA79001 were shown by two NASA scientists, Donald Bogard and Pratt Johnson [4], to be a close match to the Viking data. Further detailed studies of trapped gas components in

FIGURE 15.2 NASA's Curiosity Rover on the surface of Mars, with a cliff named Mont Merou in the background. We have studied Mars in great detail with satellites, landers, and rovers, but haven't yet returned a sample from the "red planet". However, thanks to Martian meteorites we are able to investigate the composition of Mars in the laboratory. (Image: NASA/JPL-Caltech/MSS.)

FIGURE 15.3 A view of Mars taken from the Viking 2 Lander. Viking 1 and 2 landed on Mars in 1976 and took measurements of the Martian atmosphere. Some years later, it was realised that trapped noble gases in Martian meteorites had the same composition as the atmosphere of Mars [4]. (Image: NASA/JPL-Caltech.)

other Martian meteorites have also shown a good match to the Viking atmospheric data [1]. The NASA Curiosity rover has repeated some of the measurements made by the Viking landers. Happily and reassuringly, both sets of results are in good agreement [5]. It is fair to say that the trapped gas evidence remains the most solid argument in favour of a Martian origin for these meteorites.

Additional evidence in favour of the SNCs being pieces of Mars relates to aspects of their chemical composition that seems to require formation within a large planet-sized object, rather than a relatively small asteroid [1]. So, how strong is the evidence in favour of SNCs coming from Mars? In a review article in 2000, Allan Treiman, James Gleason, and Don Bogard summarised things as follows: "*There seems little likelihood that the SNCs are not from Mars. If they were from another planetary body, it would have to be substantially identical to Mars as it now is understood*" [6].

A further important feature of Martian meteorites is their unique oxygen isotope compositions. It was first demonstrated in 1983 by Robert Clayton and Toshiko Mayeda of the University of Chicago that the Martian meteorites have an oxygen isotope composition that is distinct from that of the Earth, Moon, (4) Vesta, and most other groups of meteorites [7,8]. In a similar way to other large bodies in the inner Solar System, Mars is thought to have undergone an early phase of extensive melting [9]. This is sometimes referred to as a magma ocean stage. During this process, the planet separated into an outer silicate crust, underlying silicate mantle, and a dense, iron-rich core. Such extensive melting would have led to considerable mixing, with the result that the original oxygen isotopic variation present in its rocks would have been smoothed out. The fact that the Martian meteorites show relatively restricted oxygen isotope variation, when compared to chondrites, is clear evidence that they come from a source body that underwent global melting. In conjunction with the other lines of evidence, the oxygen isotope variation is consistent with their source body being Mars.

The combination of a distinct composition, and limited variation, made oxygen isotopes a very useful way to fingerprint Martian meteorites. But that all changed in 2013, when a meteorite designated Northwest Africa (NWA) 7034 was recovered from the North African desert [10]. NWA 7034 is a complex mixture of fragments, technically known as a breccia. These individual pieces of rock have diverse compositions (Figure 15.4) and include some that seem to be of sedimentary origin [11]. One particular feature of NWA 7034 seemed very puzzling. Its oxygen isotope composition was distinct from that of all the other recognised Martian meteorites. Initially, this made many scientists question its Martian origin. However, analysis of the noble gases: helium, neon, argon, krypton and xenon, showed that there was trapped Martian atmosphere in NWA 7034 [12]. There was also data from satellites in orbit around Mars that suggested NWA 7034 was compositionally similar to the upper layers of the Martian crust [10]. This brings us back to its weird oxygen isotope composition. In fact, it had already been demonstrated that the atmosphere of Mars and the rocky components in Martian meteorites were isotopically distinct [13]. The most likely explanation for the distinct oxygen isotope composition of NWA 7034 is

FIGURE 15.4 Martian breccia meteorite NWA 7034, sometimes nicknamed "Black Beauty". This meteorite is of Martian origin, but has an oxygen isotope composition distinct from other Martian meteorites. (Photo: Carl Agee University of New Mexico.)

that it has experienced a more significant degree of interaction with the atmosphere of Mars than other Martian meteorites. This is consistent with it being a surface-related sample from Mars.

And that brings us to the subject of returning Martian samples directly from Mars. It is not going to be easy, but happily the process is already underway. At Jezero crater on Mars, the NASA Perseverance rover is currently collecting samples of Martian rocks and soils, and placing them in sample tubes, ready for return to Earth (Figure 15.5) [14,15]. Between 21 December 2022 and 28 January 2023, the

FIGURE 15.5 The NASA Perseverance Rover is collecting samples at Jezero crater on Mars and placing them in metal tubes on the surface, ready for collection by a later mission. This image taken by the rover shows the first sample tube that was deposited on the Martian surface in December 2022. (Image: NASA/JPL-Caltech/MSSS.)

FIGURE 15.6 Getting samples back from Mars will be a complex operation, which is being done jointly by NASA and the European Space Agency (ESA). The proposed date for returning the first batch of Martian samples to Earth is 2033. In the meantime, analysis of NWA 7034, and other Martian meteorites, will undoubtedly continue to provide new and exciting information about the geology of Mars. (Image: NASA/ESA.)

rover deposited ten tubes on the Martian surface in a location called "Three Forks" [15]. But how do we get them back to Earth? Not surprisingly, this is a far from straightforward operation. However, as we saw in the previous chapter, NASA and the European Space Agency (ESA) are collaborating on a programme of missions which will ultimately lead to the return to Earth of the Perseverance samples (Figure 15.6). The planned return date is 2033. However, before that happens, Martian meteorites, including NWA 7034 and related types, will provide scientists with a wealth of new information about Mars. For example, dating of NWA 7034 has shown that it contains the most ancient material recovered from Mars to date, with ages up to 4,476 million year old [9]. It also records younger ages of between 1,500–1,200 and 135–113 million years [16], which may represent reheating and fragmentation events. Martian meteorites still have many hidden secrets to reveal about the geological evolution of Mars.

FIGURE 15.7 Recovered in 2000 by a Japanese Antarctic search team, Yamato 000593 is a large (13.8 kg) Martian meteorite, similar in composition to the historic Nakhla fall. The specimen is very fresh, with a striking green interior colouration due to its high content of the mineral pyroxene. A well-developed fusion crust is present on about 60% of the outer surface of the sample. As we have seen, this forms due to melting as the meteorite descends through the atmosphere at very high speed. The fusion crust shows well-developed flow lines that formed as the thin melted layer on the outside of the meteorite was swept off the back of the stone on its descent to Earth. (see Appendix 1 for further details.) (Image: the author.)

The fact that we already have samples from Mars, thanks to meteorites is awesome (Figure 15.7). But it gets better. As we will learn in Chapter 16, meteorites contain tiny particles that are older than our Solar System, and came from stars that have long since disappeared from our skies.

16 Grains from before the Dawn of Time

Meteorites are nice old rocks. They tell us about how our Solar System formed. Lovely! But here's the thing. They have more to say than just that. Locked inside some meteorites are grains that formed before the Solar System even existed. Grains that tell us about the stars that lived and died before the Sun first ignited. Now we are really moving into uncharted territory.

As we have seen, meteorites are complex rocks and don't give up their secrets easily. Nothing illustrates this better than the decades-long struggle to wrestle from them the secrets they held about the time before our Solar System formed. This is the story of how the tenacity and dedication of a group of international scientists extracted grains from meteorites that are literally older than the Sun.

When we looked at the Allende meteorite (Chapter 11), we saw that it contains calcium-aluminium-rich inclusions (CAIs). These high-temperature solids formed at the birth of the Solar System in a process that remains poorly understood and is still a very active area of research and debate [1]. However, most scientists agree that CAIs were probably formed close to the early Sun and most likely during energetic outbursts, which are typical of such young stellar objects (Figure 16.1) [2]. When scientists first started to understand the significance of CAIs, they got a bit carried away. They started to think that maybe everything, or nearly everything, in the Solar System formed from a very hot cloud of gas [3,4].

It was suggested that as this hot gas cloud began to cool, solid minerals formed directly from it. That is what happens when pressures are low and no liquid phase is present. This process is called condensation and it became a big buzzword in the 1970s and 1980s. Everything about the Solar System's early history was seen through the prism of condensation theory [5]. It was, to be fair, a big breakthrough. But like most revolutions, it probably went a bit too far.

A hot cloud of gas had big implications for the composition of the Solar System. In such an energetic environment, everything gets mixed up and homogenised, and any initial differences between mineral grains should be obliterated. If correct, it would have been the biggest use of a reset button in galactic history (a bit of exaggeration

DOI: 10.1201/9781003174868-16

FIGURE 16.1 NASA Spitzer Space Telescope image of a star-forming region in the vicinity of the bright, young star IRAS 13481-6124 (upper left). This young star is twenty times more massive than the Sun. The image clearly shows how the hot central stars in this region interact with the surrounding dust. The fact that some material gets heated to high temperatures seems almost certain. On the other hand, material at a distance from the hot central object probably escapes significant heating. It's a dynamic, complex environment. (Image: NASA/JPL-Caltech/ESO/Univ. of Michigan.)

there!). The Solar System would have begun life as a clean sheet of paper, isotopically homogeneous and perhaps a bit boring. The result of such a reset is that any history of matter before the formation of the Solar System would have been lost. And so, despite about 9 billion years of history from the Big Bang to the start of our Solar System, all that cosmic information would have been obliterated. Well, that's what the CAI evidence seemed to be saying.

But some scientists were not so sure. Yes, perhaps big parts of the early Solar System had been wiped clean. The birth of our Solar System, would have been a violent and energetic event (Figure 16.1). But wiped totally clean! That's asking a lot. In the 1960s, scientists studying the noble gases neon (Ne) and xenon (Xe) in carbonaceous chondrites started getting some strange results, or anomalies as they are generally called [6,7]. The variation in these elements didn't make sense in terms of being totally wiped clean. The data suggested that there seemed to be small grains in these meteorites that had compositions well outside the normal variation observed in most samples [6]. Perhaps some material that was older than the Solar System had survived after all. It was a deeply subversive idea. But science and scientists are like that. Getting them to follow a party line is a bit like herding cats. Forget it! That's just not going to happen.

Case in point, Don Clayton (Figure 16.2), an astrophysicist and cosmologist who just didn't believe in the total wipe clean story at all [8]. Yes, parts of the Solar System would have got very hot indeed, but not all. Far out in the dusty outer reaches, very old grains would remain largely unaltered, before becoming trapped into asteroids and then meteorites. All you had to do was find them. And the reward for all this hard work was that these grains would tell us about the years and years before the Solar System began. It was the ultimate prize in the study of meteorites, and neon and xenon were showing the way [6,7]. The hunt was definitely on [9].

There were many great scientists chasing this prize. One in particular was Edward Anders of the University of Chicago (Figure 16.2). For a while, Ed worked with Colin Pillinger's group at the Open University. Colin's team had pioneered a technique that took a residue of the meteorite, in which these presolar grains were concentrated, and combusted them at increasingly higher temperatures [10]. This technique is known as stepped combustion. This allowed them to analyse the composition of the grains and to get clues about what they were made of. Based on this work, it started to become clear that carbon was important. Carbon was present in

FIGURE 16.2 This photo was taken at a NASA workshop held in November 1990. Don Clayton is standing on the right of the photo holding a glass of water. Immediately to the right of him (standing) is Ernst Zinner, and immediately to the left (sitting) is Edward Anders. (Photo: AIP Emilio Segrè Visual Archives, Clayton Collection.)

two forms: graphite (Figure 16.3) and diamond [11]. Another important presolar mineral was silicon carbide (Figure 16.4) [11].

However, the stepped combustion technique had a number of significant drawbacks. Firstly, it used up large amounts of meteorite. Perhaps 2 or 3 g had to be dissolved in very strong acids to remove 99% of the material present, just to make a concentrate of the more robust presolar grains. Secondly, any weaker, more fragile presolar grains were lost. And thirdly, as most of the meteorite itself was destroyed, you had no idea where the presolar grains were located, e.g., in CAIs, chondrules, or matrix.

But things were moving fast. Huge advances were being made in the development of an analytical instrument known as a secondary ion mass spectrometer, or SIMS for short (Figure 16.5) [13]. This could take the meteorite sample, and with only a relatively limited amount of preparation (it didn't need to be dissolved in acid), allow you to hunt and then analyse presolar grains in situ. It had many of the characteristics of an electron microscope, but with the advantage that you could make isotopic measurements on the presolar grains. It was a significant advance on the step combustion technique.

One of the great pioneers of SIMS analysis of presolar grains was Ernst Zinner [14]. He was based at Washington University, St. Louis when making his ground-breaking discoveries. At a distance, I saw him in action before the start of the 1991 Meteoritical Society held in Monterey, California. On the way to the conference with Robert Hutchison, we stopped in St. Louis to visit his former PhD student,

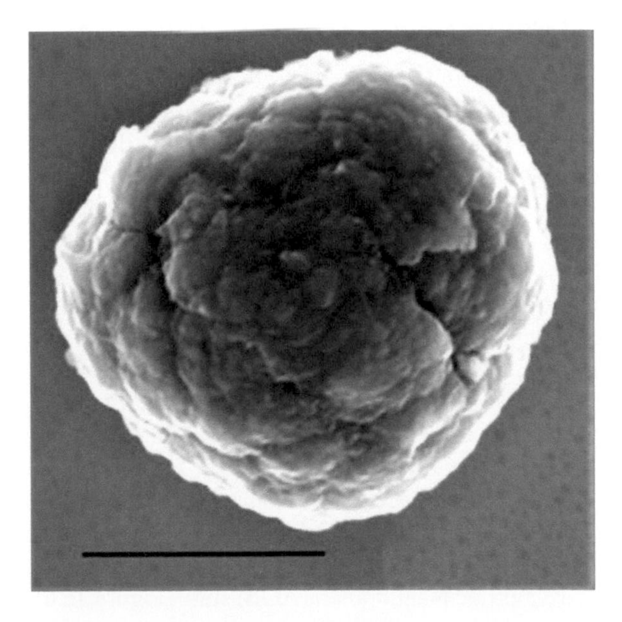

FIGURE 16.3 A presolar graphite grain from the Murchison meteorite. The scale bar is 1 μm (one-thousandth of a millimetre). These grains likely formed when stars that were much larger than the Sun reached the end of their lives, and having exhausted all their fuel, exploded as supernovae. (Image: Philipp Heck.)

FIGURE 16.4 An image of a presolar silicon carbide grain from the Murchison meteorite taken using a scanning electron microscope. Presolar silicon carbide grains formed predominantly in asymptotic giant branch (AGB) stars [11]. These are stars similar in size to our Sun which have exhausted their nuclear fuel (Figure 16.6). This particular grain is approximately 8 μm (eight thousandths of a millimetre) in its longest dimension. (Image modified from Heck et al. [12].) (Image: Philipp Heck.)

Conel Alexander, who was also involved in presolar grain research. At the time, this research felt like the holy grail of meteorite studies, and still does. Using meteorites to study the period before the Sun even existed is incredible [11]. In the years since the pioneering work of Anders, Zinner, and others, the subject has come on apace [7,9,11,15]. In addition to diamond, graphite, and silicon carbide, SIMS analysis has been used to identify presolar silicon- and oxygen-rich grains (silicates), as well as various oxides and silicon nitride Si_3N_4 [7,11].

Presolar grain studies clearly indicated that these materials were not formed in our Solar System [7,11,15]. So where did they come from?

The most likely place to generate them was in the stars that lived and died before our Sun existed [11]. In a very real sense, we are dealing with stardust. Grains from ancient stars that could be studied in the laboratory, to reveal the secrets of a time before the Sun was born. But the evidence from presolar grains suggested that they could not have formed from a single source. At least two distinct environments were needed. The presolar silicon carbides seemed to have been produced in stellar objects known as asymptotic giant branch (AGB) stars (Figure 16.6), whereas presolar carbon and graphite were principally formed in supernovae (Figure 16.7) [11,15]. AGB stars are similar in size to the Sun, but they have reached the end of their lives, having exhausted almost all of their nuclear fuel. In this final stage of their life cycle, they send out large amounts of material into interstellar space and eventually form

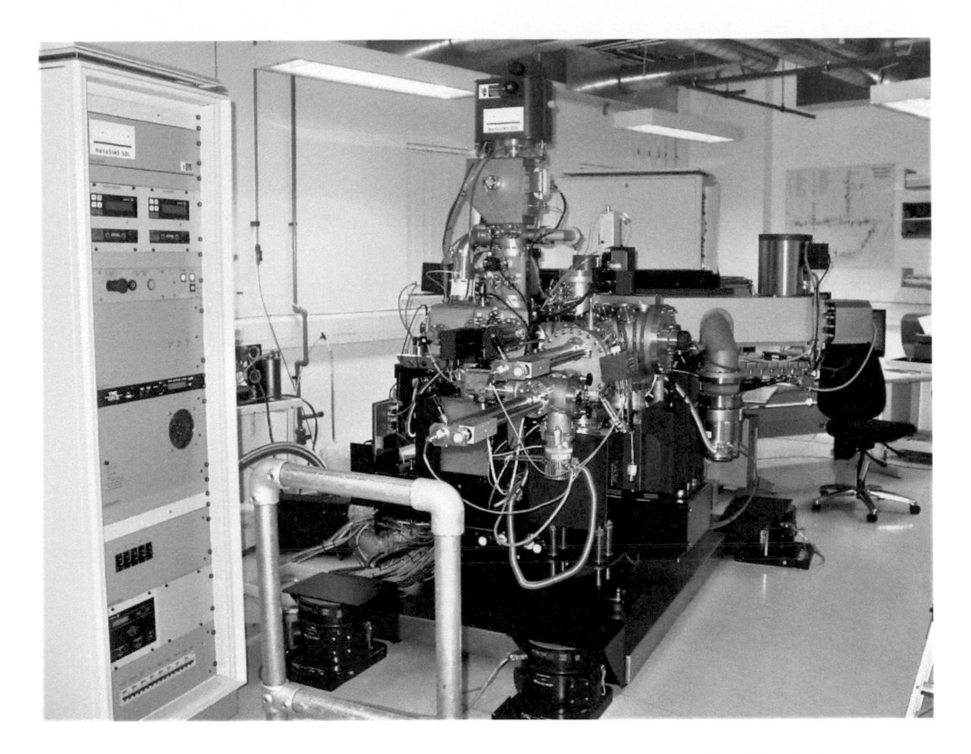

FIGURE 16.5 The Open University's Cameca NanoSIMS. This is effectively a powerful microscope that can image presolar grains and then analyse their isotopic composition. (Photo: the author.)

what is known as a planetary nebula (Figure 16.6). Supernovae, on the other hand, are cosmic explosions that take place when a star tears itself apart [16].

Stars are the engines in which elements are synthesised [17]. Only the very lightest elements, hydrogen, some helium, and a little bit of lithium, were produced in the Big Bang event, 13.8 billion years ago, from which our universe is thought to have formed. All the other elements are produced in stars. As you might expect, it's not a simple process [17]. The elements that can be made in a star depend on its size. The heavier elements, up to iron, only form in more massive stars, at least eight times bigger than the Sun [11]. And it is these larger stars that end their lives as supernovae [16]. These massive stellar explosions turn out to be crucial in the formation of the very heaviest elements. These include the precious elements such as gold, silver, and platinum, as well as the more industrially useful ones such as copper, zinc, and tungsten. At the same time as the supernova is throwing out these newly formed heavy elements, it also ejects into interstellar space all the other elements that were previously formed in the star [17]. This material is then recycled into the next generation of stars. Our Sun is not a first generation star, far from it. Presolar stardust grains bear witness to the fact that generations of earlier stars lived out their lives and died prior to the formation of the Solar System. Ultimately, it is those long dead massive stars we have to thank for all the gold that is now present on the Earth.

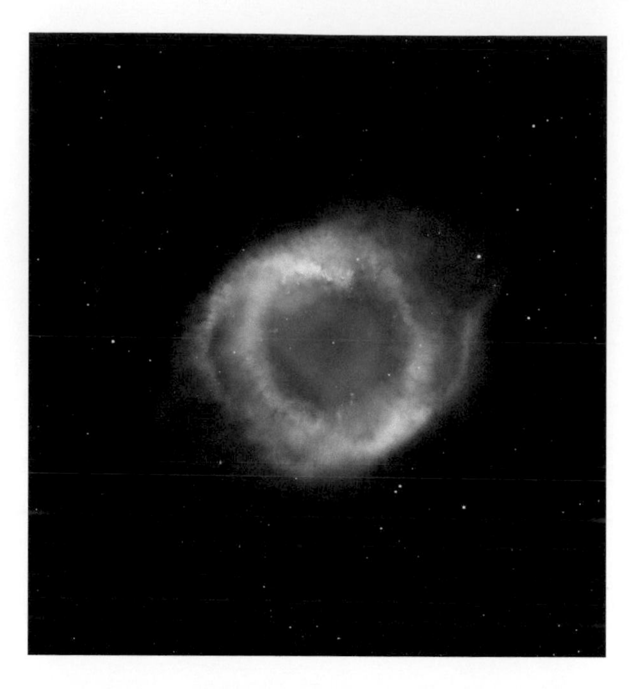

FIGURE 16.6 NGC 7293 Helix Nebula is an example of a planetary nebulae which formed from an asymptotic giant branch (AGB) star. AGB stars are Sun-like stars at the end of their lives, which have exhausted all their hydrogen and helium fuel. Our Sun will go through an AGB stage in about 5,000 million years' time and end as a planetary nebula similar to NGC 7293. Presolar silicon carbide grains are predominantly formed in AGB stars. (Image: Spitzer Space Telescope – NASA/JPL-Caltech.)

And presolar grain research continues to produce some astonishing results. A team, led by Philipp Heck of the Field Museum of Natural History and the University of Chicago, dated 40 large presolar silicon carbide grains from the Murchison meteorite and found that some had ages predating our Solar System by as much as 3,000 million years [12]. Since our Solar System is generally accepted to be nearly 4,600 million years old, that could make some of the grains 7,600 million years old or more (the error bars on these dates are fairly large). If that conclusion wasn't astonishing enough, there's more!

The silicon carbide grains dated in this study, as we have seen, likely formed in the outflows of AGB stars. When they get to the AGB stage, these ageing stars shed huge amounts of material into the interstellar medium. There it would have wafted around for a while, before being trapped into the collapsing gas and dust from which our own Solar System formed. Philipp Heck and co-authors suggest that most of these grains were probably only free floating in space for a few hundred million years prior to the formation of our Solar System. The grains are thus able to provide us with an insight into the time gap between the end phase of one generation of stars and the birth of the next. A few hundred million years may sound like a long time, but compared to the age of our Solar System (4,567 million years), it is quite short.

FIGURE 16.7 The Crab Nebula was formed by a supernova event that probably took place in 1054 AD. In July or August of that year, Chinese astronomers observed a new bright star that was even visible in the daytime sky. However, like all supernovae, it would have faded within a few months of its initial appearance. Supernovae are important events for a number of reasons. They spread the elements formed in stars into deep space, enriching the gas and dust that is already present there and from which new stars will ultimately form. In addition to the elements already formed in the giant star before it explodes, the supernova itself forms new heavy elements, such as gold, silver, platinum, and uranium. (Image: NASA, ESA, J. Hester and A. Loll (Arizona State University).)

And what of the future? Meteorites provide us with samples from the very dawn of the Solar System, and contain material that is even older than that. Being able to interrogate such material in the laboratory provides us with the opportunity to extract precious information about how we came to be here. It is a challenge to decipher the limited clues that remain following the long and eventful history these samples have experienced. But the prize is worth it. Or to sum it up in another way, meteorites are cool. But having got this far, you already knew that. Thanks for reading.

Frequently Asked Questions

1. How can you tell a meteorite from a normal rock?

Well, it's not usually a big problem when the meteorite is fresh. Due to frictional heating, as it travels at high speed through the atmosphere, the outer surface of a meteorite melts. This melted material is swept off the back of the meteorite during most of its flight, but as the object slows, it freezes onto its outer surfaces, forming a thin, black layer known as "fusion crust" (Figure A1.1).

FIGURE A1.1 Fusion crust is a diagnostic feature of meteorites. This is Yamato 75102, collected by a Japanese team in the Antarctic Yamato Mountains in 1975. The original weight of the specimen was 11 kg, so a large specimen for Antarctica. It is an ordinary chondrite (L4/5), with a light coloured interior and dark fusion crust on the outside. The fusion crust is a very thin layer, no more than 1 mm in thickness, that coats the outside of the specimen. (Photo: the author.)

But things can be a bit trickier for meteorite “finds”; these are samples that were not observed to fall and may have been out in all weathers for a very long time before an individual piece was recovered. Not surprisingly, finds can be quite weathered. Appendix 1 provides some useful tips on what to look out for when trying to tell a space rock from an earthling.

2. Where are the best places to find meteorites?

Unfortunately, cool temperate regions in the northern hemisphere are not such good places to find meteorites, mainly because there are plenty of terrestrial rocks lying about to confuse things. Even if you find a piece of metal-rich material, it is unlikely to have come from space, and more likely to be related to our long industrial heritage. It may be a piece of slag (see Appendix 1). The best places to search for meteorites turn out to be deserts, both hot and cold (see Chapters 12 and 13).

3. How old are meteorites?

Meteorites are the oldest rocks in the Solar System. In fact, a distinct type of object found in meteorites, known as calcium-aluminium-rich inclusions, or CAIs for short, are used to date the age of the Solar System at 4,567 million years old. Chapters 11 and 16 provide further details.

4. What is a meteorite?

A meteorite is a piece of natural space debris that survives its entry through the atmosphere and lands on the Earth.

5. Where do meteorites come from?

Most meteorites are fragments from small asteroids found in the asteroid belt located between the orbits of Mars and Jupiter (Chapters 7 and 10). However, we also have meteorites that came from the Moon and Mars (Chapters 13 and 15).

6. What are meteorites made of?

In terms of their silicate-rich minerals, meteorites are not so different from rocks that you would find on the Earth. However, the most common type, the ordinary chondrites, contains metal, which is something you don’t normally find in terrestrial rocks. The metal content of some meteorites can be very high. The variability in metal content displayed by meteorites is used as a simple way to classify them into three types: stones, irons, and stony-irons. Appendix 2 provides a short guide to meteorite classification. An extended version of these meteorite classification notes is available on my website: Meteorites: The Blog from the Final Frontier https://meteoritestheblog.com

7. How often do meteorites land on the Earth?

Meteorites arrive on Earth on a regular basis, but most go unwitnessed and either drop into the sea or on sparsely populated areas of land. However, on average, between about five to ten meteorites per year are witnessed during their atmospheric entry, with material subsequently recovered from the Earth's surface. These are known as meteorite "Falls".

Estimates of how much space material arrives on the Earth vary wildly, but could be as much as 60,000 metric tons per year (Chapter 3).

8. How big are they?

Meteorites come in a wide range of sizes, from micrometeorites that are only a fraction of a millimetre in diameter (Chapter 3), to metre-sized specimens. The largest intact meteorite is the Hoba meteorite, which was found in Namibia in 1920, and measures approximately 1 m × 3 m × 3 m and weighs about 60 tons [1]. The Chelyabinsk meteorite that exploded over Russia in February 2013 is estimated to have been 20 m or so in diameter (Chapter 4). Meteorites can get even bigger. The dinosaur killer is estimated to have been 10 km in diameter (Chapter 4).

9. How much could you expect to pay for a meteorite?

Meteorites are rare and precious objects, so they are collected and traded by both enthusiasts and professional dealers. At the low end, one could pay a few dollars per gram for a weathered ordinary chondrite. At the other end of the scale, a meteorite from the Moon or Mars might fetch as much as $2,000 per gram. The trade in meteorites is discussed in Chapters 1 and 13.

10. What's the risk that the planet Earth will be wiped out by a meteorite?

As we saw in Chapter 6, planet Earth formed by collisions between small asteroids at the birth of the Solar System. However, there is now no chance that an asteroid, even a giant one, could destroy the Earth, it is just too big. Large asteroids of similar size to the one that killed the dinosaurs (~10 km in diameter) have only hit the Earth on rare occasions in the last few billion years (Chapter 4). Another such mega-impact would have huge consequences for the biosphere. It is estimated that the dinosaur-killing asteroid wiped out 75% of all life on the Earth. But the Earth itself can easily withstand such an impact and given time the biosphere would recover too. Unfortunately, humanity would be unlikely to survive such a catastrophe.

11. Did the dinosaurs really get wiped out by a meteorite?

There seems little doubt that a major mass extinction at the Cretaceous-Tertiary boundary 66 million years ago was caused by a meteorite impact into the area now occupied by the Yucatan peninsula in Mexico. The buried impact structure, known as the Chicxulub crater, is about 180 km in diameter (Chapter 4).

12. What's the difference between an asteroid and a meteorite?

An asteroid is a rocky object that orbits in space, whereas a meteorite is a piece of space debris that survives atmospheric entry and lands on the Earth. In 2008, a 4-m-diameter asteroid 2008 TC3 was tracked before entry and exploded over Sudan, and pieces that were recovered were given the name Almahata Sitta. So, in space it was asteroid 2008 TC3 but when material was recovered from the ground it was a meteorite called Almahata Sitta [2].

13. What's the difference between a shooting star and a meteorite?

A shooting star is another name for a meteor. These are small grains of space dust that burn up in the upper atmosphere at heights of between about 80 and 120 km above the Earth [3]. In contrast, a meteorite is a piece of natural material that makes it through the atmosphere and lands on the Earth's surface.

14. What is the difference between a meteor and a meteorite?

As meteors and shooting stars are the same thing, the answer to question 13 applies equally well to this question.

15. What can we learn from meteorites?

Meteorites provide us with a wealth of knowledge about the origin and evolution of the Solar System. Most are close to 4,567 million years old (give or take a few million years) and so are the oldest rocks you will ever come across. As discussed in Chapter 16, some meteorites even contain dust that predates the Solar System.

16. Have traces of life ever been found in a meteorite?

This is a very interesting and important question. Claims were made that the Martian meteorite ALH 84001 contained features that were interpreted as being fossilised life forms (Chapter 12). It was suggested that these provided evidence in favour of life on Mars. Unfortunately, this evidence has not stood up to close scientific scrutiny and is no longer accepted by the majority of scientists [4,5]. In Chapter 11, we saw that some carbonaceous chondrites contain amino acids. But these comprise both the left and righthanded varieties in equal amounts. This is the signature of chemical synthesis rather than any involvement by living organisms. So, for the moment at least, there is no evidence for alien life in meteorites.

17. What percentage of meteorites that land on the Earth are recovered?

Unfortunately, it's only a tiny percentage. You need the right conditions to find meteorites. Hot and cold deserts are best. In Chapter 3, we looked at micrometeorites and saw that these represent the vast bulk of extraterrestrial material arriving on the Earth. However, recovering micrometeorites is not easy and only a miniscule percentage of this size fraction is collected each year.

18. Where are the best places to see meteorites?

National science museums and galleries are really the best places to get up close to meteorites. These treasure houses exhibit amazing extraterrestrial space rocks and there are often specimens that you can touch, which is nice. They will also generally have plenty of modern interactive displays, full of important information about meteorites and asteroids.

19. Do meteorites fall on the Moon and other planets?

Yes, there are named meteorites brought back from the Moon and official meteorites located on the surface of Mars by rovers [6]. The Meteoritical Bulletin Database

FIGURE A1.2 This large iron meteorite was identified by the NASA Curiosity Rover on the surface of Mars. Strictly speaking it is two iron meteorites, the big piece (2 m from end to end) at the back was given the informal name "Lebanon" and the small, detached piece at the front was called "Lebanon B". Data collected by Curiosity was able to demonstrate to the satisfaction of the NomCom that these were iron meteorites and so they have been given the official names Lebanon = Aeolis Palus 002 and Lebanon B = Aeolis Palus 003. (Image: NASA.)

currently lists 15 meteorites that have been located on the surface of Mars and 2 meteorites that were returned from the Moon by the Apollo astronauts (Figure A1.2).

20. Do meteorites hit satellites?

Yes, this can be a big problem. Satellites in orbit around the Earth are regularly hit by micrometeorites. In some cases, these encounters can cause the satellite to malfunction [7]. It is possible to take avoidance measures to protect satellites when the Earth is passing through a known meteor shower. Engineers operating the James Webb Space Telescope (JWST) turn the spacecraft optics away from the direction of the shower in advance of its arrival [8]. Unfortunately, not all large impact events are predictable and in May 2022 the JWST sustained a significant micrometeorite impact on one of its primary mirror segments. Although there was a slight loss of data quality, the telescope is designed to withstand such events and its operating performance remains well above requirements [8].

21. How many people are killed by meteorites each year?

None, as far as we know. In Chapters 1 and 4, we examined the potential risks from meteorites. While meteorites pose a very low risk to human life, the threat level is not zero. Events such as the Tunguska explosion in Siberia in 1908, and the Chelyabinsk meteorite strike in 2013, demonstrate that there is potential for a large loss of life from meteorite impacts. In recognition of these dangers, NASA undertook the Double Asteroid Redirection Test (DART) [9]. On 26 September 2022, the DART spacecraft impacted into the surface of Dimorphos, a small asteroid that orbits a larger asteroid Didymos. The test was a big success, significantly altering the orbit of Dimorphos. These encouraging results hint at the possibility that an asteroid heading towards the Earth could potentially be deflected from its collisions course.

NOTES FOR FAQS

[1] *Hoba Meteorite.* Wikipedia (accessed 01 January 2024). https://en.wikipedia.org/wiki/Hoba_meteorite.

[2] Official Meteoritical Bulletin Database Entry for Almahata Sitta (accessed 01 January 2024). https://www.lpi.usra.edu/meteor/metbull.php?code=48915.

[3] Meteor FAQs, American Meteor Society. https://www.amsmeteors.org/meteor-showers/meteor-faq/#2.

[4] Joseph Brean (2022) *Not Proof of Life after All: 'Fossil' in Famous Martian Meteorite Made by Water, Not Aliens.* National Post (accessed 01 January 2024). https://nationalpost.com/news/world/not-proof-of-life-after-all-fossil-in-famous-martian-meteorite-made-by-water-not-aliens.

[5] A. Steele et al. (2022) Organic synthesis associated with serpentinization and carbonation on early Mars. *Science*, **375**, 172–177. https://www.science.org/doi/10.1126/science.abg7905.

[6] Linda Martel (2017) *Bounty of Iron Meteorites Found on Mars.* CosmoSparks, PSRD (accessed 01 January 2023). http://www.psrd.hawaii.edu/CosmoSparks/Nov17/irons-on-mars.html.

[7] Colin Schultz (2013) *How Do You Shield Astronauts and Satellites From Deadly Micrometeorites?* Smithsonian Magazine (accessed 01 January 2023). https://www.smithsonianmag.com/smart-news/how-do-you-shield-astronauts-and-satellites-from-deadly-micrometeorites-3911799/.

[8] Thaddus Cesari (2022) *Webb: Engineered to Endure Micrometeoroid Impacts.* NASA James Webb Space Telescope (accessed 01 January 2023). https://blogs.nasa.gov/webb/2022/06/08/webb-engineered-to-endure-micrometeoroid-impacts/.

[9] NASA DART. *Double Asteroid Redirection Test. Mission Overview* (accessed 01 January 2024). https://dart.jhuapl.edu/Mission/index.php.

Appendix 1
So, You Think You Have Found a Meteorite – What Next?

It happens! That dark, black rock that you are 100% certain wasn't there the last time you looked. And it's got such a weird shape. And then there's all those shiny metal bits all over it. What could it be? You pick it up. Wow! It's really heavy. You give it a quick sniff. OK, so nothing unusual there then. But what on Earth is it? It would certainly make a good paperweight. You take it indoors. The Grangers are coming around for supper tonight. And so, after a nice glass of Chardonnay, or two, the conversation moves on to movies. Everyone thought Leonardo DiCaprio was brilliant in that weird end of it all, asteroid, comet thingy. What was it called now? Don't look about? Don't look over? Don't look up? Don't look up – that was it! Who did you say was the new Dr Who? Well, they were good in I'm a celebrity! Then you remember that rock from the garden. It gets passed around. The word "meteorite" jumps into the conversation. Could it be a meteorite? Well, it does look a bit alien, and it is heavy. Someone suggests that a good test would be to see if it is magnetic. A magnet is duly found and yes! It is very magnetic. And so that's it. The strange rock becomes a suspect meteorite. But what to do next, how can you get it verified? You know that the chances of it being a bona fide space rock are very, very low, but still, what if it is really the genuine article? A small piece from another world. You can't just ignore that can you?

And of course, you shouldn't.

Space rocks are rare, extremely rare, but if everyone decides not to take things further when they locate a suspicious-looking rock, well, science would be much the poorer. To help, here is a quick guide to some of the things you need to look for. It's not exhaustive, but it will give you a better idea whether your sample is a genuine extraterrestrial, or just an unusual earthling.

THINGS TO LOOK OUT FOR

Features That Are Diagnostic of Meteorites

1. **Fusion crust:** This is always the most diagnostic feature for identifying fresh meteorites. All newly fallen meteorites will have a thin layer of fusion crust on their outer surfaces. It does vary somewhat in colour and texture between different meteorite types. Fusion crust is a very thin, dark, usually glassy, coating on the outside of a fresh meteorite. It forms during the final stages of atmospheric flight. A meteorite is heated to very high temperatures as it descends through the atmosphere, and the outside of the object is melted and vapourised. As the fragment decelerates, it cools and the temperature on its outer surface drops accordingly. Finally, the last remaining dregs of liquid coating the exterior solidify, forming a layer that is no thicker than 1 mm. It is also a fragile layer, so often if the fragment was bashed by something during the final stages of its fall, the fusion crust in places may become detached. That's good because it gives you the opportunity to see how thin the fusion crust layer is. Also, the interior of the meteorite will most likely be light in colour in contrast to the fusion crust (see Figure A1.1).

FIGURE A1.1 Small fragment from the 2013 Chelyabinsk meteorite fall. This sample is only about 2 cm long and is almost totally enclosed in dark fusion crust. In a number of places, the fusion crust layer has become detached, and you can clearly see the light-coloured interior. (Scale: between longer divisions, 5 mm). (Photo: Andy Tindle.)

FIGURE A1.2 Gibeon iron meteorite showing well-developed regmaglypts or "thumb-print" structures. (Photo: the author.)

2. **Regmaglypts:** Scientists just love complicated words for what are really quite simple ideas, and this is another example. Regmaglypts are shallow depressions or dimples on the surface of a meteorite. Some people call them "thumb-print" structures, which is what they look like. On their own, they are not a particularly diagnostic feature because lots of terrestrial rocks can show a similar texture. But in certain cases, they can be very distinctive and may be a helpful pointer to identifying a space rock (Figure A1.2).
3. **Fusion crust flow lines:** As a meteorite moves through the atmosphere, the molten liquid on its surface is pushed backwards and off the rear of the flying stone by its movement through the air. When the fusion crust freezes, thin veinlets of once molten liquid are preserved. It is a delicate texture, but absolutely diagnostic (Figure A1.3).
4. **An orientated shape:** It has to be remembered that a meteorite flying through the atmosphere is losing material all the time. This process is known as ablation. If the stone is twisting and tumbling as it moves, there will be an even amount of ablation across its outer surface. But sometimes a flying fragment will maintain a single orientation, a bit like a rocket capsule returning to Earth. In this case, the meteorite fragment will take on an orientated shape with a conical front and a rounded back. There may also be flow lines across the surface of the orientated face. These features are rare, but very beautiful (Figure A1.4).

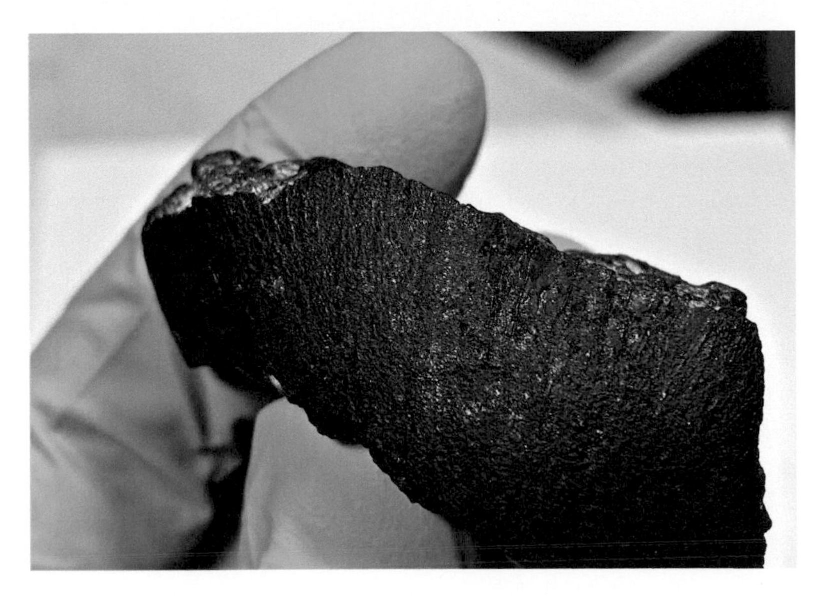

FIGURE A1.3 If you look carefully at the fusion crust on this piece on the Chelyabinsk meteorite, you can see that it shows flow lines running vertically in this photo. (Photo: the author.)

FIGURE A1.4 This meteorite from the Japanese National Polar Research Institute collection (NIPR) shows an orientated shape. This developed in the atmosphere when the stone was undergoing ablation. However, rather than tumbling as is more usual, this particular specimen maintained a fixed orientation, a bit like a rocket's heat shield. (Photo: the author.)

5. **A light-coloured interior:** As discussed, the fusion crust is fragile and may become dislodged. The interior of most meteorites is relatively light in colour and this will be clearly seen where the fusion crust has fallen off. This is an interesting characteristic, which only works for some meteorite types, and not for others. The most abundant group of meteorites arriving on Earth are ordinary chondrites, and in general, they have an interior that is light in colour compared to fusion crust. This is also the case for a number of other types of meteorites, such as the eucrites. But for one important group, known as carbonaceous chondrites, it doesn't work because their interiors are just as dark as the fusion crust (Figure 10.8). Iron meteorites also do not show any significant difference between the colour of their exteriors and interiors (Figure A1.5).
6. **Chondrules:** Perhaps you have a hand lens or magnifying glass handy. If so, it would be worth looking in a bit more detail at the structure of any interior areas that are exposed through the fusion crust. Chondrules are up to millimetre-sized spherical objects present in chondrites (Figure A1.6), which include the most common type of meteorites, the ordinary chondrites.

FIGURE A1.5 A relatively large and broken piece of the Chelyabinsk meteorite surrounded by small, fully fusion-crusted Chelyabinsk "peas". The outer surface of the larger piece is also coated in fusion crust. The interior is mainly light in colour, but is crossed by darker shock veins. It is clear in a number of places that the shock veins are truncated by the outer surface of the stone. The shock veins formed when the meteorite was part of a much large asteroid which collided with another object in the asteroid belt. (Photo: the author.)

FIGURE A1.6 This piece of the Allende (CV3) carbonaceous chondrite contains abundant, often very large chondrules. Many are several millimetres in diameter. If you can see them, chondrules are an excellent feature that confirms you have found a meteorite. There are all sizes in this meteorite, and if you look carefully, they are everywhere. Chondrules are even more abundant in ordinary chondrites. Allende also contains large calcium aluminium-rich inclusion (CAIs). These are the white, irregularly shaped, objects up to a centimetre or so in long dimension. There is a big CAI at the centre of the field of view. (Photo: the author.)

7. **Metal:** Most meteorites have a highly variable content of metal. Iron meteorites are naturally very rich in metal. Taken together with other features discussed above, the presence of metal in a sample may be helpful in identifying it as a meteorite. But many industrial waste materials contain metal, so the presence of metal is not always the most diagnostic feature.

Features That Are Not So Helpful

Set out below are some common features that people think are diagnostic of meteorites, but are not:

1. **Weight:** This is an interesting one. Yes, iron meteorites are very heavy, meaning that they have a much higher density than terrestrial rocks. But most meteorites have a similar density to rocks from Earth, so for a similar size they have much the same weight. And some meteorites actually feel quite light when you pick them up, i.e., have a relatively low density. If your sample is very heavy, does that make it an iron meteorite? Well, unfortunately not. One of the materials most often mistaken for being a meteorite is waste from industrial metal production. This is sometimes referred to as "slag". It comes in an almost limitless variety of textures and colours. It also has very variable densities. When you pick it up, it can be very heavy and metallic in appearance, hence can be easily mistaken for an iron meteorite.

In summary, while weight is something that can make a specimen stand out, it is not a particularly diagnostic characteristic.

2. **Magnetic characteristics:** Most of the common meteorite types do show magnetic characteristics and will deflect a magnet suspended on a piece of string. But many terrestrial rocks contain the mineral magnetite and so also show magnetic properties. Industrial waste materials (slag) are also often magnetic. This means that the magnetic properties of a sample are not really helpful. And there is a downside too. If your sample is a meteorite and you wave a magnet close to it, you will almost certainly disturb its natural remnant magnetism. That's a pity because the natural magnetism in a meteorite has the potential to reveal a lot of information about conditions in the early Solar System. The moral here is to keep your magnet in your pocket. It won't really help, and if you are lucky and have found a meteorite, then demagnetising it with a hand magnet is a bit sad considering it may have been magnetised for the last 4.5 billion years, and you have just put an end to it.
3. **Colour and external texture:** We have already looked at fusion crust and seen that it is a very diagnostic characteristic of meteorites. But there are many terrestrial rocks that have fine-grained smooth exteriors that superficially look like fusion crust. You need to look carefully at the exterior of your sample Figure A1.7. Is the dark coating very thin, or does it persist all the way through the specimen? Is the coating a single thin layer that wraps around the specimen, or is it of highly variable thickness and shows changes in colour and texture? Real fusion crust is often very uniform in colour and is a very thin layer.

FIGURE A1.7 Your sample may have a smooth dark exterior and may even look as though it has regmaglypts. You may be convinced it is a space rock. But is that shiny surface very thin and fragile, or is it an external polish? The specimen shown in the photo is not a meteorite. It is not coated in fusion crust and the surface depressions are not regmaglypts. It is an example of a "meteorwrong". (Photo: the author.)

"Meteor Wrongs!" – Things That Are Often Mistaken for Meteorites

Identifying a real meteorite is not easy. Remember they are very rare objects. Unfortunately, there are a number of commonly found types of material that can be mistaken for meteorites. Here are a few examples:

1. **Industrial waste materials – "slag":** By far and away, the "stuff" most commonly mistaken as being a meteorite is industrial waste material, otherwise known as "slag". It is just everywhere. Humans have been smelting metals since about 5,000 BC. That's a long time! Until very recently, there have been no restrictions on where slag can be deposited. Even now, it is commonly used for a wide variety of purposes: in the construction industry, to repair driveways, fill potholes, etc. You get the picture. And don't think that because an area looks green and wooded that it was always like that. When a medieval smelting works was abandoned, it would quickly revert back to being fully forested. Unfortunately, this means that slag is found in all sorts of non-industrial looking areas. Slag can often show many of the characteristics we expect of a meteorite: dark, fine-grained, heavy, magnetic. It can, and often does, contain native metal. Slag is the curse of the meteorite hunter, and some familiarity with this material will help you to identify real space rocks (Figure A1.8).
2. **Concretions:** Many sedimentary rocks, particularly limestone, contain dark, iron-rich spherical objects called concretions. They bear a superficial resemblance to some meteorites, and so can be misidentified as space rocks. However, the resemblance is superficial, and if you study the images of real meteorites given in this book, that will help to tell them apart (Figure A1.9).

FIGURE A1.8 Industrial waste from metal smelting can sometimes be mistaken for space rocks. The fragments in the picture are not extraterrestrials, just slag. They may look a bit like meteorites, but on closer scrutiny they are clearly not the real deal. Unfortunately, slag is literally everywhere. (Photo: the author.)

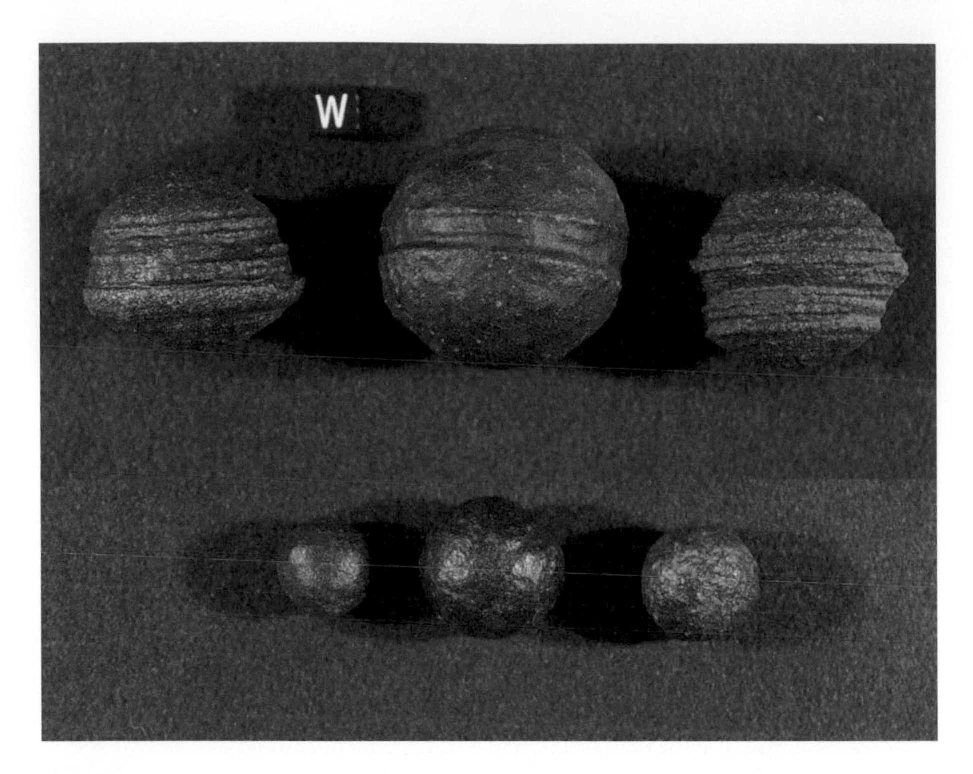

FIGURE A1.9 Spherical, iron-rich concretions are a common feature of sedimentary rocks. They can look a bit like the general idea of what a meteorite ought to look like. But if you study concretions carefully, they don't really show the typical characteristics of space rocks discussed earlier. The examples shown in the image are known as Moqui Marbles, and represent hematite concretions, from the Navajo Sandstone of Utah. Scale cube, 1 cm across. (Photo: Wikipedia/Paul Heinrich).

What If I Still Think My Sample Is a Meteorite?

As you would expect, a little bit of experience does help when trying to identify space rocks and even the experts get it wrong from time to time. A great source of information when trying to identify a meteorite is the website maintained by Professor Randy Korotev of Washington University, St. Louis, USA: https://sites.wustl.edu/meteoritesite/identification/

It is also worth remembering that a suspected meteorite is almost never identified on the basis of its physical appearance alone. That is just the start of the process. Before the NomCom gives it an official name, the mineral composition of any sample must be determined using instruments such as an analytical scanning electron microscope.

Finally, if you are still convinced you have found a meteorite, your best next step is to contact an expert. The way you do that will depend on where you live. A local university with a geology department might be a good place to start. If you look on the web, there will be contacts you can find in your area who will be able to provide some further advice. You could also reach out to museums which have a space rock display. It can take a bit of time and persistence to get an expert to look at your sample, but you will eventually succeed. And don't be too disappointed if they say it is not a meteorite. As mentioned already, they are rare objects, and many terrestrial rocks are mistaken for meteorites before a real one turns up. But here is the good news, they do turn up eventually. It is worth remembering that the great meteorite hunter Harvey Nininger was told by a "meteorite expert" at the start of his career not to bother looking for space rocks, because it would take him forever to find one. But Nininger completely ignored this advice and went on to find literally hundreds of meteorites.

Appendix 2
Meteorites – A Very Short Guide

Here is a short guide to the main types of meteorites that should help you to pick your way through the minefield of extraterrestrial jargon. An extended version of these classification notes is available on my website: *Meteorites: The Blog from the Final Frontier* https://meteoritestheblog.com

In this book, I have tried to keep technical jargon to a minimum. Getting overly obsessive with the buzzwords puts many people off and makes the subject seem dense and unexciting, which it definitely is not. However, that said, it is always useful to have a few definitions so we can all talk from the same page. And it is also important to understand some of the details of meteorite classification to get a better feel for the rich variety of materials that arrive on Earth from space. Meteorites are not all the same, far from it.

With these noble aims in mind, we will first define a few key terms and then look at the main types of meteorites.

A FEW DEFINITIONS

First up, let's start with the word **meteorite** itself. Basically, a meteorite is the rock you pick up from the ground, not the bit that flies through the air. To paraphrase one widely used definition of the term "meteorite": it is a space rock that survives its passage through the atmosphere, and so can be picked up off the ground [1,2]. On the other hand, as it flies through the atmosphere doing its stuff, it is more technically referred to as a **meteor** [1,2]. Now this can be a bit puzzling, because when most meteorites arrive, the term "meteor" is not usually used to describe their atmospheric flight. Instead, the term most often used is **fireball**. But it turns out that this is not really as confusing as it might seem. A fireball is just a bright meteor [1]. You kind of know it when you see it. More strictly speaking, a fireball is a meteor that is brighter than any planet or star [1]. The brightness is given a magnitude number, but we won't worry about that here [1]. But it doesn't quite end there. Some fireballs are very big and as a result you get a bit more stuff going on, explosions, shock waves, etc. These larger fireballs are sometimes referred to as **bolides** [1,3]. A bolide is "*a very bright meteor which fragments or explodes. Sounds of the explosion can be heard if the observer is close enough*"

[1]. But it can get even more exciting, because you can also have a **superbolide** [3]. These are rare, spectacular bolides and have a strict definition in terms of the amount of light given out, which again we don't really need to worry about. The Tunguska and Chelyabinsk events were examples of superbolides. There probably should be a category called megabolides for the sort of dinosaur-killing, giant-sized, objects that have happily only turned up extremely rarely in Earth's history. But as it's very unlikely any of us would survive one of them to compare notes afterwards, that's probably not a useful term.

We have been talking a lot about **asteroids** in the book. These are basically chunks of space rock that hurtle around the Solar System. It's a very useful term so we use it a lot. The term **meteoroids** has been introduced for mini-asteroids, which range from the size of a grain of sand up to 1 m in diameter [2,3]. Asteroids, meteoroids, meteors, and meteorites – that's starting to get a bit confusing. So, let's think about a small chunk of rock hurtling through space. Away from the Earth, it's an asteroid (or meteoroid if smaller than 1 m). As it descends through the atmosphere, it's a meteor (or fireball if bright enough). Finally, when it gets picked up off the ground, it's a meteorite. In the main part of this book, I have tried to keep life a little simpler and talk about asteroids in space and meteorites (and space rocks) when they land on Earth. I apologise if the term meteoroid has not been used as often as it should.

A few other useful terms now. Many incoming space rocks fragment as they reach the lower, denser parts of the atmosphere. When those fragments fall to Earth, they define a **strewn field**. This is an elliptical area elongated in the flight direction of the meteorite. When you look closely at the broken surface of many meteorites, you may be able to make out millimetre-sized spherical structures, which are called **chondrules**. These important objects formed very early in Solar System history (Chapter 2) [4]. They are widely believed to have started out as clumps of dust that were flash-heated, and so melted to form hot, free-floating spheres of liquid rock. They come in an amazing variety of textures and nothing like them has been formed since. Then there are **calcium-aluminium-rich inclusions**, or **CAIs** for short (Chapter 11). An awful name for astonishingly important objects. CAIs are rich in the elements calcium and aluminium, as you would expect, but more importantly, they are the oldest Solar System objects so far dated. In fact, CAIs are used to date the formation of our Solar System. The current most reliable date is 4,567 million years ago, which is pretty old.

DIFFERENT TYPES OF METEORITES

An important note before we get going. It is much easier to make sense of meteorites if you can see what they look like. A great way to do this is by visiting **the Virtual Microscope** and taking a look at the wonderful images it contains. All are the work of Andy Tindle. The site contains superb, high-quality, rotatable images of a wide variety meteorite types. Here are the main areas to visit:

British and Irish meteorites

https://www.virtualmicroscope.org/content/british-irish-meteorites

This has not only thin-section images, but also rotating object images.

EUROPLANET meteorites:

https://www.virtualmicroscope.org/content/europlanet-meteorites
Includes thin-section images of some important types of achondrites.

Lunar meteorites:

https://www.virtualmicroscope.org/content/lunar-meteorites
Just one sample at the moment, but a growing collection.

Martian meteorites:

https://www.virtualmicroscope.org/content/martian-meteorites
Lots of thin-section images. Check out the rotating Nakhla specimen.

Apollo Lunar Missions:

https://www.virtualmicroscope.org/collections/apollo
A collaboration with NASA. If you want to see Moon rocks, this is the place.

Naturally, any individual meteorite that lands on Earth, and is then recovered for scientific study, represents a unique and important sample. But scientists have been studying space rocks for well over 200 years [5], and in that time, it has been recognised that differences and similarities exist between the various samples we have in our worldwide collections. Finding connections between samples and trying to work out which samples are related to each other is a natural human process. We do it all the time. We try to bring some sort of order to what appears to be a chaotic world. If supermarket shelves were randomly filled with unrelated objects, we would pretty soon start complaining. We expect things to be ordered in some sort of logical way. It's the same with space rocks. Set out below is a very much simplified scheme for dividing up the various types of meteorites. For those who would like a more in-depth treatment, there are some very authoritative scientific papers [6] and websites [7] that can help. An extended version of these classification notes, is available online at: *https://meteoritestheblog.com/meteorites-a-very-short-guide/*

Stones, Irons, and Sony Irons

A simple but effective way to divide up meteorites is on the basis of how much metal compared to silicate (rocky) material they contain. This leads to three main divisions: (1) **stones**, which are predominantly composed of silicate-rich minerals, and in some respects, resemble rocks found on the surface of the Earth; (2) **stony-irons**, which are, more or less, 50:50 mixtures of silicate-rich minerals and nickel-bearing metal; and (3) **irons**, which are essentially composed of nickel-bearing metal, but also contain various other minor minerals.

Stones – The Power of Chondrules

In terms of number of samples, stones are by far the most important type of meteorites. But here is the catch! The fact that meteorites are made up predominantly of rocky material is not really very helpful when trying to understand the origin of these samples. It turns out chondrules are the key. As we discussed in Chapter 2, these small silicate spheres, generally no more than a few millimetres in diameter, likely formed in the first few million years of Solar System history. They are usually easy to spot so that helps us to divide up the stony meteorite world into two major types. If a meteorite has chondrules, it is called a "chondrite", and if it lacks chondrules, it is an "achondrite". Phew! We have made a start. Let's look at chondrites first.

Chondrites

Ok, so you might think that's it. It's got chondrules, so it is a chondrite. It turns out that's not a bad way to look at things. But there are some complexities. Aren't there always!

Most chondrites are stuffed full of chondrules. In fact, some types have almost nothing else but chondrules. However, there are other chondrites in which chondrules are much rarer and are separated from each other by fine-grained material. We call that fine-grained stuff, matrix. Some of these matrix-rich chondrites are super dark and there's a reason for that: they have a high content of carbon. And there are other differences. When you rotate the broken surface of a fresh chondrite, it will often glint. That is because it contains nickel-bearing metal. The amount of metal in a chondrite also helps to define which particular group a chondritic meteorite falls into.

So, you can see that chondrites have a range of features that can be used to subdivide them. This is not an exact science and there are sometimes disagreements about what features are the most important. But broadly, chondrites are divided into three major types (often called classes), two further minor types, and then a bunch of samples that don't seem to fit in anywhere (they are called ungrouped).

Major types (classes) of chondrites:

1. Ordinary chondrites
2. Carbonaceous chondrites
3. Enstatite chondrites

Minor types: Rumututi-type (R) chondrites and Kakangari-type (K) chondrites and ungrouped chondrites.

We could look in great detail at the important features of each of these different types of chondrites. It is a fascinating subject, but there is a lot of information already out there [6,7]. The principal features of each of the three main types of chondrites are set out below.

Ordinary Chondrites

First of all, this is not a very nice name. They are called "ordinary" not because they are boring, but because they are the most common type of meteorite that falls to Earth. At the time of writing (February 2024), the Meteoritical Bulletin Database [8] lists the total number of classified and officially approved meteorite falls as being 1,238. Of this number, 968 (78.2%) are ordinary chondrites. This compares with only 52 (4.2%) carbonaceous chondrite falls, and 17 (1.4%) enstatite chondrite falls. All other meteorite falls total 201 (16.2%). These numbers clearly show how dominant ordinary chondrites are in our meteorite collections. If finds are included, the dominance of ordinary chondrites increases slightly, such that they represent 84.5% of all meteorites that have ever been collected.

Characteristically ordinary chondrites are densely packed with chondrules. They contain very little fine-grained matrix and almost no calcium-aluminium-rich inclusions. They also have variable content of metal. Now we don't really need to go into lots of detail on this, but the amount of iron they contain helps to subdivide the ordinary chondrites. Three groups are recognised: H group, which stands for high iron, L which stands for low iron, and LL which stands for low metal and low iron.

Carbonaceous Chondrites

This grouping covers a very diverse range of meteorites. You might think from the name they all contain a lot of carbon. Most do, but some don't. Not much of a help then. They contain a lower fraction of chondrules than ordinary chondrites, and a lot more of that fine-grained stuff we call matrix. Many of them contain abundant CAIs. There are some fairly coarse-grained types with big CAIs, such as Allende (Chapter 11), and others which are highly altered, such as the CIs we discussed in Chapter 14. You are probably wondering how these meteorites are actually classified. Well, you have to use a range of characteristics, not just one. Their chemistry is important. They have a lot of calcium and aluminium compared to silicon. Their oxygen isotopes are also diagnostic. Eight main groups are recognized (CV, CR, CH, CI, CO, CB, CM, and CK), and there are also some smaller groupings that we don't need to worry about.

Enstatite Chondrites

In a similar way to ordinary chondrites, enstatite chondrites often have lots of chondrules and can also have a lot of metal. The mineralogy of enstatite chondrites is important in classifying them; they are dominated by a type of mineral known as low-calcium pyroxene, rather than olivine, as is found in ordinary chondrites. Once again, their oxygen isotope compositions are also a big help.

Achondrites

As their name implies, these meteorites do not contain chondrules. But apart from that, most of the different groups in this category have little in common with each other. Achondrites include planetary types derived from the Moon (Chapters 12 and 13), Mars (Chapters 12 and 15), and Vesta (Chapter 7), the second largest asteroid in the asteroid belt. There is an enigmatic group of very ancient achondrites that shares many characteristics with the planetary achondrites, except no one knows where they are from. These are the angrites. There is a group of achondrites that seem to be related to the enstatite chondrites. These are the aubrites. A number of achondrite groups share some chemical characteristics with the chondrites and are sometimes referred to as primitive achondrites. Again, these are a very diverse group and include brachinites, acapulcoite-lodranites, winonaites, and ureilites. There are also an increasing number of so-called ungrouped achondrites. These are an area of active research because they may sample long destroyed parent asteroids that formed very early in Solar System history.

Stony-Irons

There are only two main types of stony-irons: pallasites and mesosiderites. Mesosiderites are a mix of crustal rocks that resemble the HED meteorites with metal-rich material (Chapter 8). Pallasites consist of a mix of coarse-grained olivine and metal.

Irons

Irons are classified based on the variation of a number of key elements, including nickel, gallium, iridium, and germanium (Chapter 5). Using this data, irons are divided into 12–14 principal groups. There are also a lot of ungrouped irons. It is estimated that iron meteorites are likely derived from about sixty parent asteroids.

NOTES FOR APPENDIX 2

[1] American Meteor Society – Glossary. https://www.amsmeteors.org/resources/glossary

[2] What is the difference between meteors, meteorites and meteoroids? For some straightforward answers to these questions this NASA website is helpful: https://science.nasa.gov/solar-system/meteors-meteorites/
Amore rigorous, scientific look at this subject is given by Rubin and Grossman (2010): Alan E. Rubin and Jeffrey N. Grossman (2010) Meteorite and meteoroid: New comprehensive definitions. *Meteoritics and Planetary Science*, **45**, 114–122. https://onlinelibrary.wiley.com/doi/full/10.1111/j.1945-5100.2009.01009.x

[3] The International Astronomical Union (IAU) provides the following guidance concerning the definition of meteor and meteorite-related terms:
(i) Meteors & Meteorites: The IAU Definitions of Meteor Terms. https://www.iau.org/public/themes/meteors_and_meteorites/
(ii) Definitions of terms in meteor astronomy. https://www.iau.org/static/science/scientific_bodies/commissions/f1/meteordefinitions_approved.pdf

[4] James N. Connelly, Matin Bizzarro, Alexander N. Krot, Ake Nordlund, Daniel Wieland and Marina A. Ivanova (2012) The Absolute Chronology and Thermal Processing of Solids in the Solar Protoplanetary Disk. *Science*, **338**, 651–655.

[5] Ursula B. Marvin (2006) Meteorites in history: an overview from the Renaissance to the 20th century in McCall, G.J.H., Bowden, A.J. & Howarth, R.J. (eds) 2006. The History of Meteoritics and Key Meteorite Collections: Fireballs, Falls and Finds. *Geological Society, London, Special Publications*, **256**, 15–71.

[6] Michael K. Weisberg, Timothy J. McCoy and Alexander N. Krot (2006) Systematics and Evaluation of Meteorite Classification. In: Lauretta, Dante S. and Sween, H. Y., Jr., (eds) *Meteorites and the Early Solar System II*. University of Arizona Press, pp. 19–52. https://repository.si.edu/bitstream/handle/10088/20577/min_Weisberg_et_al_2006_In_Meteorites_and_the_Early_Solar_System_II_Univ_of_Arizona_Press_19-52.pdf?sequence=1&isAllowed=y

[7] Kerry Lotzof, Types of meteorites, Natural History Museum (29 accessed December 2023). https://www.nhm.ac.uk/discover/types-of-meteorites.html#:~:text=There%20are%20three%20main%20types,which%20mostly%20have%20silicate%20minerals

[8] Meteoritical Bulletin Database (accessed 29 December 2023). https://www.lpi.usra.edu/meteor/

Chapter Notes

A live version of these notes and links is available from the folllowing website: https://meteoritestheblog.com/notes-and-links/

CHAPTER 1

[1] In the world of science, before you can publish anything about a meteorite it must have an official name. This process is overseen by a single committee of the Meteoritical Society, known as the Nomenclature Committee. It is the sole global organisation that has the authority to authenticate an individual meteorite and all scientists and reputable dealers take its deliberations very seriously. Once the NomCom, as it is generally referred to, gives a new sample an official name an entry is created in the Meteoritical Bulletin. It still comes out on a regular basis as a printed journal, but all official entries also appear pretty much instantaneously on the NomCom's official database, which is generally referred to as the Met Bull Database. The whole operation was created by Dr Jeff Grossman of NASA, and is a major tool for meteorite researchers everywhere. Check out the entry for Valera to get a feel for how it works: https://www.lpi.usra.edu/meteor/metbull.php?code=24149 (accessed 27 December 2023).

[2] Fall and Finds. These may sound like trivial terms, but they are not. A meteorite "fall" is a witnessed event following which a meteorite sample is collected. There must be a reasonable level of certainty that the samples of the meteorite collected on the ground were linked to the events that have been witnessed. As discussed in later parts of the book, there may be a series of phenomenon associated with a fall event, such as a bright fireball, explosions, infrasound etc. In contrast, a "find" is just that, a meteorite that is located and authenticated but was not observed arriving on Earth.

[3] Ordinary chondrites are the most common type of meteorites arriving on Earth. There is more detail about them in Appendix 2. They are subdivided into three groups: L, LL, and H. The number 5 tells you how much the meteorite was heated within its parent asteroid. A type 5 was heated quite a lot. An L5 is a very common type of meteorite.

[4] M. Horejsi (2006) The Accretion Desk, Valera Revisited. Meteorite Times. https://www.meteorite-times.com/Back_Links/2006/August/Accretion_Desk.htm (accessed 27 December 2023).

[5] Sale Details for Valera Meteorite (2016) Christie's Auction House, London, UK. https://www.christies.com/en/lot/lot-5987687 (accessed 27 December 2023).

[6] Details for Valera Meteorite (2021) Christie's Auction House, New York, USA, Sale. https://onlineonly.christies.com/s/deep-impact-martian-lunar-other-rare-meteorites/valera-only-documented-death-meteorite-24/112846 (accessed 27 December 2023).

[7] D. Steele (2002) Rocks on Your Head. The Guardian Newspaper, 17th January 2002. https://www.theguardian.com/science/2002/jan/17/technology1 (accessed 27 December 2023).

[8] A. A. Childs (2011) One Hundred Years Ago Today, a Mars Meteorite Fell in a Blaze. Smithsonian Magazine. https://www.smithsonianmag.com/smithsonian-institution/one-hundred-years-ago-today-a-mars-meteorite-fell-in-a-blaze-23722099/

[9] Nakhla 1, Collection Martian Meteorites, The Virtual Microscope. https://www.virtualmicroscope.org/content/nakhla-1 (accessed 27 December 2023).

[10] M. Horejsi (2011) New Concord Meteorite – Let's Stop Kicking a Dead Horse. Meteorite Times Magazine. https://www.meteorite-times.com/the-new-concord-meteorite-lets-stop-kicking-a-dead-horse/ (accessed 27 December 2023).

[11] Met Bull Database Entry for Vaca Muerta meteorite. https://www.lpi.usra.edu/meteor/metbull.php?code=24142 (accessed 27 December 2023).

[12] At a simple level, meteorites can be divided into three types: stones, irons, and stony-irons. Vaca Muerta is a stony-iron type in which there is roughly a 50:50 mixture of metal and rocky materials. The metal is mainly composed of iron, but also contains a lot of nickel. In contrast, Valera is a stone and contains much less metal than Vaca Muerta.

[13] The total weight of Valera is 50 kg, and although it broke into three pieces on impact, the object that hit the cow would likely have been a single mass as it descended earthwards. Or would it? Meteorites often fragment during atmospheric entry and arrive at the Earth's surface as a "shower" of stones. There could have been other pieces of the Valera meteorite that were never recovered, but that is just speculation. Ordinary chondrites have a density of about 3 grams per cubic centimeter. Assuming a spherical geometry for Valera that density indicates that the mass of rock that hit the cow was a boulder-sized object with a diameter of about 32 cm.

[14] Fireball FAQs. The American Meteor Society. https://www.amsmeteors.org/fireballs/faqf/ (accessed 27 December 2023).

[15] The celebrated Myth Busters cannonball incident took place in 2011 and received wide media coverage at the time. Here are some links to various online resources which give further details. Associated Press Video (2011) "Mythbusters" Cannonball Hits Family Home, Van. https://www.youtube.com/watch?v=BJY45bADSqQ (accessed 27 December 2023). Linda Holmes (2011) So the Mythbusters Punched a Hole in a House with a Cannonball. Now What? National Public Radio (npr). https://www.npr.org/2011/12/08/143349193/so-the-mythbusters-punched-a-hole-in-a-house-with-a-cannonball-now-what (accessed 27 December 2023).

[16] Cannonballs of old, as fired from classic, heavy cannons are more technically a type of "round shot" e.g., Thomas Flynn (2011) Shot (including musket balls, cannon balls and bullet moulds). Portable Antiquities Scheme, county pages, finds recording pages, time periods, post medieval. https://finds.org.uk/counties/findsrecordingguides/shot/ (accessed 27 December 2023).

[17] E. Pierazzo and H. J. Melosh (2000) Understanding oblique impacts from experiments, observations, and modeling. *Annual Review of Earth and Planetary Sciences*, **28**, 141–167. https://www.annualreviews.org/doi/abs/10.1146/annurev.earth.28.1.141.

[18] E. Wright et al. (2020) Ricochets on asteroids: experimental study of low velocity grazing impacts into granular media. *Icarus*, **351**, 113963. https://doi.org/10.1016/j.icarus.2020.113963; https://arxiv.org/pdf/2002.01468.pdf

[19] Food and Agriculture Organization of the United Nations - Livestock System – Cattle (2024). https://www.fao.org/livestock-systems/global-distributions/cattle/en/

[20] Sarah Catherine Walpole et al. (2012) The weight of nations: an estimation of adult human biomass. *BMC Public Health*, **12**, 439. https://bmcpublichealth.biomedcentral.com/articles/10.1186/1471-2458-12-439

[21] The most common type of cattle reared in Venezuela are Criollos, otherwise known as Spanish Longhorns: Venezuelan Criollo, Wikipedia. https://en.wikipedia.org/wiki/Venezuelan_Criollo (accessed 27 December 2023).

[22] United States Department of Agriculture - Foreign Agricultural Service - Global Agricultural Information Service (GAIN) (2019) Livestock and Products Annual, Venezuela - Report No. VE2019-0001. https://apps.fas.usda.gov/newgainapi/api/Report/DownloadReportByFileName?fileName=Livestock%20and%20Products%20Annual_Caracas_Venezuela_08-15-2019 (accessed 27 December 2023).

CHAPTER 2

[1] R. Hutchison, J. C. Barton and C. T. Pillinger (1991) The L6 chondrite fall at Glatton, England, 1991 May 5. *Meteoritics*, **26**, 349. https://articles.adsabs.harvard.edu/pdf/1991Metic..26S.349H

[2] Jonathan Morris (2012) Devon 'meteorite' brings memories of close encounter. BBC New website https://www.bbc.co.uk/news/uk-england-20002161. NOTE: As Glatton fell a little before the internet era had really kicked off, there are only a few bits of information about its arrival on the web. The article above by BBC Plymouth News in 2012 features the reflections of their reporter who was based in Cambridgeshire at the time that the Glatton meteorite landed in Mr Pettifor's garden. It is a generally accurate account of what took place. However, the suggestion is made that Mr Pettifor kept the meteorite at his home wrapped in cling film for several months. But this is incorrect. The meteorite was passed to the Natural History Museum shortly after the fall for safe keeping. The meteorite did go back to the village for the local fete a little after its arrival on Earth. But only for a couple of days. That could have been when it got wrapped in cling film.

[3] Jerome Real (2014) Parts Departing from Aircraft (PDA), Safety First #18. https://mms-safetyfirst.s3.eu-west-3.amazonaws.com/pdf/safety+first/parts-departing-from-aircraft-pda.pdf (accessed 27 December 2023).

[4] Airportwatch website (2015) 2 ft Diameter Metal Diffuser Fell from Plane Near Chicago onto a Water Park (Nobody Hurt). https://www.airportwatch.org.uk/2015/11/2-ft-diameter-metal-diffuser-fell-from-plane-near-chicago-onto-a-water-park-nobody-hurt/ (accessed 27 December 2023).

[5] BBC News website (2018) "Airline Poo" Falls on India Village Causing Confusion. https://www.bbc.co.uk/news/world-asia-india-42772918 (accessed 27 December 2023).

[6] Glatton, Virtual Microscope, Collection: British & Irish Meteorites. https://www.virtualmicroscope.org/content/glatton. You can also view the Glatton meteorite in all its glory by visiting the Virtual Microscope website and trying out for yourself an "object rotation" of the specimen. https://www.virtualmicroscope.org/sites/default/files/html5Assets/glatton_o/index2.html?specimen=/node/307. The Glatton sample is covered in a very thin dark rind of fusion crust. This forms due to frictional heating with the atmosphere. The temperatures reached at the surface of the descending meteorite are high enough to melt and vaporise the outer layers of the stone. But only the very outer part. As the rock turns to liquid and vapour, it is swept off the back of the hurtling stone. This produces a glowing dust and vapour trail. This process of heating and

removal of the hot products is known as ablation and has the huge advantage that only the outer part of the stone is heated significantly. The ablation process acts like the heat shield on a returning spacecraft. Finally, as the flying stone slows down, the ablation process comes to an end and any remaining liquid on the outside cools to form a thin glassy, black layer, generally no more than 1 mm thick. The Glatton specimen has been sampled and pieces removed for scientific analysis. This is why some parts are light in colour as they represent in the interior of the meteorite. When it first landed the Glatton stone was totally enclosed in black fusion crust.

[7] As it penetrates deeper into the atmosphere, the forces on a meteorite build up very rapidly, particularly at the front end. When these forces exceed the internal strength of the space rock, it often breaks up catastrophically into a mass of smaller fragments. These then land close to each other and the whole group of stones is called a meteorite shower.

[8] Nick Catford and Bob Jenner (2003) South Kensington Home Security Region 5 War Room. Subterranea Britannica. https://www.subbrit.org.uk/sites/south-kensington-home-security-region-5-war-room/ (accessed 27 December 2023).

[9] Natural History Museum, London Human Remains Collection. https://www.nhm.ac.uk/our-science/collections/palaeontology-collections/london-human-remains-collection.html (accessed 27 December 2023).

[10] Jay Sullivan 14 highlights from the spirit collection The Natural History Museum. https://www.nhm.ac.uk/discover/14-must-see-spirit-collection.html (accessed 27 December 2023).

[11] I. Mansfield (2015) Holborn Tube Station's Unexpected Gift to Particle Physics Research. IanVisits. https://www.ianvisits.co.uk/articles/holborn-tube-stations-unexpected-gift-to-particle-physics-research-13924/ (accessed 27 December 2023).

[12] Iconic British 1960s Science Fiction Film: Quatermass and the Pit. https://en.wikipedia.org/wiki/Quatermass_and_the_Pit_(film) (accessed 27 December 2023).

[13] The authoritative record of the Glatton meteorite is given on the Met. Bull Database. https://www.lpi.usra.edu/meteor/metbull.php?code=10930 (accessed 27 December 2023).

[14] An excellent review of what chondrules are and how they might have formed: Harold C. Connolly Jr. and Rhian H. Jones (2016) Chondrules: The canonical and noncanonical views. *Journal of Geophysical Research: Planets*, **121**, 1885–1899. https://agupubs.onlinelibrary.wiley.com/doi/full/10.1002/2016JE005113

[15] Rhian H. Jones (2017) New Group Paper about the Origin of Chondrules. Earth and Solar System. https://earthandsolarsystem.wordpress.com/2017/05/03/new-group-paper-about-the-origin-of-chondrules/ (accessed 27 December 2023).

[16] Martin Bizzarro, James N. Connelly and Alexander N. Krot (2017) Chondrules: Ubiquitous Chondritic Solids Tracking the Evolution of the Solar Protoplanetary Disk. In: Pessah, M., Gressel, O. (eds) *Formation, Evolution, and Dynamics of Young Solar Systems*. Astrophysics and Space Science Library, **445**, Springer, Cham. https://doi.org/10.1007/978-3-319-60609-5_6

[17] D. Rowe, Henry Clifton Sorby. Yorkshire Philosophical Society. https://www.ypsyork.org/resources/yorkshire-scientists-and-innovators/henry-clifton-sorby/ (accessed 27 December 2023).

[18] Noel Chaumard, Munir Humayun, Brigette Zanda and Roger H. Hewins (2018) Cooling rates of type I chondrules from Renazzo: implications for chondrule formation. *Meteoritics and Planetary Science*, **53**, 984–1005. https://doi.org/10.1111/maps.13040

[19] H. Kaneko, K. Sato, C. Ikeda and T. Nakamoto (2023) Cooling rates of chondrules after lightning discharge in solid-rich environments. *The Astrophysical Journal*, **947**, 15. https://iopscience.iop.org/article/10.3847/1538-4357/acb20e

CHAPTER 3

[1] J. Pelley (2017) Dust unsettled. *ACS Central Science*, **3**(1), 5–9. https://pubs.acs.org/doi/10.1021/acscentsci.7b00006

[2] A. de Castro (2021) Part of the Dust Accumulating in Your House Comes from Outer Space. United Academica Magazine. https://www.ua-magazine.com/2021/04/16/part-of-the-dust-accumulating-in-your-house-comes-from-outer-space/ (accessed 27 December 2023).

[3] Annual extraterrestrial flux rates are subject to significant errors. The numbers quoted in the text (20,000 to 60,000 metric tons per year) are based on estimates from Esser and Turekian (1988) and Love and Brownlee (1993). It is also important to note that these values represent the amount of material that arrives at the top of the atmosphere. A large amount of this material will not reach the Earth's surface as distinct grains, but instead is ablated away on entry. These lost particles will still add material to the Earth, but not as recoverable grains. The amount of material that survives down to the surface is significantly less than the flux rates given by Esser and Turekian (1988) and Love and Brownlee (1993). Based on their study of cosmic dust from the Transantarctic Mountains, and estimates from other studies, Suttle and Folco (2020) suggest that the annual surface deposition rate for cosmic dust is between 1,500 and 6,500 metric tons per year. Using Suttle and Folco's upper value, we can estimate the maximum number of particles that are likely to fall each year per square metre of the Earth's surface. Love and Brownlee (1993) indicate that the largest size fraction of dust has a diameter of 0.2 mm and an estimated mass of 0.000015 g. If we assume for the purposes of calculation that all particles have this size and mass, then 6,500 metric tons (6.5×10^9 g) represent 4.3×10^{14} particles. The surface area of the Earth is approximately $5.1 \times 10^{14}\,m^2$. That gives a value of 0.8 particles per square metre. In view of the uncertainties, let's call that 1 particle per square metre per year. B. K. Esser and K. K. Turekian (1988) Accretion rate of extraterrestrial particles determined from osmium isotope systematics of Pacific pelagic clay and manganese nodules. *Geochimica et Cosmochimica Acta*, **52**, 1383–1388. https://www.sciencedirect.com/science/article/abs/pii/0016703788902098; S. G. Love and D. E. Brownlee (1993) A direct measurement of the terrestrial mass accretion rate of cosmic dust. *Science*, **262**, 550–553. https://www.science.org/doi/10.1126/science.262.5133.550; M. D. Suttle and L. Folco (2020) The extraterrestrial dust flux: size distribution and mass contribution estimates inferred from the transantarctic mountains (TAM) micrometeorite collection. *Journal of Geophysical Research: Planets*, **125**, e2019JE006241. https://doi.org/10.1029/2019JE006241

[4] Y. Wong. How Small Can the Naked Eye See? BBC Science Focus Magazine. https://www.sciencefocus.com/the-human-body/how-small-can-the-naked-eye-see/ (accessed 27 December 2023).

[5] Office for National Statistics (2020) One in Eight British Households Has No Garden. https://www.ons.gov.uk/economy/environmentalaccounts/articles/oneineightbritishhouseholdshasnogarden/2020-05-14#:~:text=The%20median%20garden%20size%20for, in%20Scotland%20(the%20largest) (accessed 27 December 2023).

[6] A. Cutmore (2022) Revealed! How Often the Average Brit Moves House IdealHome. https://www.idealhome.co.uk/news/zoopla-average-brit-moves-home-181281# (accessed 27 December 2023).

[7] The Challenger Expedition. Dive and Discover - Expeditions to the Seafloor - Woods Hole Oceanographic Institution. https://divediscover.whoi.edu/history-of-oceanography/the-challenger-expedition/ (accessed 27 December 2023).

[8] J. Woolf (2022) The Challenger Expedition: Peering into the Abyss. Royal Scottish Geographical Society. https://www.rsgs.org/blog/the-challenger-expedition-peering-into-the-abyss (accessed 27 December 2023).

[9] F. T. Kyte (2002) Tracers of the extraterrestrial component in sediments and inferences for Earth's accretion history. In Koeberl, C., and MacLeod, K.G., eds., Catastrophic Events and Mass Extinctions: Impacts and Beyond: Boulder, Colorado, *Geological Society of America Special Paper* **356**, 21–38. https://ntrs.nasa.gov/api/citations/20030062938/downloads/20030062938.pdf
[10] NASA Curation. Aircraft Collected Particles (ACP) Collection. https://curator.jsc.nasa.gov/dust/acp_collection_description.cfm (accessed 27 December 2023).
[11] NASA Curation. Cosmic Dust the NASA Cosmic Dust Collections. https://curator.jsc.nasa.gov/dust/index.cfm (accessed 27 December 2023).
[12] NASA Curation – Cosmic Dust the South Pole Water Well (SPWW). https://curator.jsc.nasa.gov/dust/spww_collection_description.cfm (accessed 27 December 2023).
[13] M. J. Genge, J. Larsen, M. Van Ginneken and M. D. Suttle (2017) An urban collection of modern-day large micrometeorites: evidence for variations in the extraterrestrial dust flux through the quaternary. *Geology*, **45**, 119–122. https://inspectapedia.com/Microscopy/Micrometeorites-Genge.pdf.
[14] H. H. Nininger (1941) Collecting small meteoritic particles. *Popular Astronomy*, **49**, 159–162.
[15] M. D. Suttle, T. Hasse and L. Hecht (2021) Evaluating urban micrometeorites as a research resource-a large population collected from a single rooftop. *Meteoritics and Planetary Science*, **56**, 1531–1555. https://onlinelibrary.wiley.com/doi/full/10.1111/maps.13712
[16] Jon Larsen (2023) "In Search of Stardust: the discovery of the fresh "urban" cosmic dust" Speaker: Jon Larsen, Project Stardust (University of Oslo, UiO). https://www.youtube.com/watch?v=y22KEbHRZRg (accessed 27 December 2023).
[17] Jon Larsen (2022) Project Stardust. https://projectstardust.xyz/micrometeorites (accessed 27 December 2023).
[18] Thilo Hasse. Searchable Urbane Mikrometeorite. https://www.micrometeorites.org/database (accessed 27 December 2023).
[19] American Meteor Society, Meteor FAQs. https://www.amsmeteors.org/meteor-showers/meteor-faq/#4 (accessed 27 December 2023).
[20] Leonids NASA Explore. https://leonid.arc.nasa.gov/meteor.html#:~:text=Size%3A%20Most%20visible%20Leonids%20are,and%20weights%20only%200.00006%20gram (accessed 27 December 2023).
[21] Jarmo Moilanen, Maria Gritsevich and Esko Lyytinen (2021) Determination of strewn fields for meteorite falls. *Monthly Notices of the Royal Astronomical Society.*, **503**, 3337–3350. https://academic.oup.com/mnras/article/503/3/3337/6155056

CHAPTER 4

[1] E. Osterloff. How an Asteroid Ended the Age of the Dinosaurs. Natural History Museum. https://www.nhm.ac.uk/discover/how-an-asteroid-caused-extinction-of-dinosaurs.html (accessed 27 December 2023).
[2] At the time that the dinosaurs went extinct, the Earth was experiencing a massive phase of volcanism in India. The products from this event are known as the Deccan Traps. This volcanism was accompanied by the release of vast amounts of CO_2 which may have raised global temperatures by 5°C. This would have resulted in significant levels of climate change. While this may not have been enough to account for the end of Cretaceous extinction event, it may have been a contributory factor. Read more about the Deccan traps here: University of California, Berkeley. The causes and impacts of Deccan volcanism at the end-Cretaceous: https://deccan.berkeley.edu/# (accessed 27 December 2023).

[3] P. Voosen (2019) Did volcanic eruptions help kill off the dinosaurs? *Science*. https://doi.org/10.1126/science.aax1020. https://www.science.org/content/article/did-volcanic-eruptions-help-kill-dinosaurs (accessed 27 December 2023).

[4] M. Wills (2017) Dinosaurs Could Have Avoided Mass Extinction If the Killer Asteroid Had Landed Almost Anywhere Else. The Conversation. https://theconversation.com/dinosaurs-could-have-avoided-mass-extinction-if-the-killer-asteroid-had-landed-almost-anywhere-else-87109 (accessed 27 December 2023).

[5] F. L. Condamine, Guillaume Guinot, Michael J. Benton and Philip J. Currie (2021) Dinosaur biodiversity declined well before the asteroid impact, influenced by ecological and environmental pressures. *Nature Communications*, **12**, 3833. https://doi.org/10.1038/s41467-021-23754-0

[6] G. R. Wieland (1925) Dinosaur extinction. *The American Naturalist*, **59**, 557–565. https://www.journals.uchicago.edu/doi/epdf/10.1086/280066

[7] R. Black (2013) The Top Ten Weirdest Dinosaur Extinction Ideas. Smithsonian Magazine. https://www.smithsonianmag.com/science-nature/the-top-ten-weirdest-dinosaur-extinction-ideas-23642539/ (accessed 27 December 2023).

[8] R. Black (2012) Disease and the Demise of the Dinosaurs. Smithsonian Magazine. https://www.smithsonianmag.com/science-nature/disease-and-the-demise-of-the-dinosaurs-122975049/ (accessed 27 December 2023).

[9] H. K. Erben, J. Hoefs and K. H. Wedepohl (1979) Paleobiological and isotopic studies of eggshells from a declining dinosaur species. *Paleobiology*, **5**, 380–414. https://www.jstor.org/stable/2400319

[10] N. C. Koch (1967) Disappearance of the dinosaurs. *Journal of Paleontology*, **41**, 970–972. https://www.jstor.org/stable/1302167

[11] J. M. Cys (1967) The inability of dinosaurs to hibernate as a possible key factor in their extinction. *Journal of Paleontology*, **41**, 266. https://www.jstor.org/stable/1301929

[12] S. Gartner and J. Keany (1978) The terminal cretaceous event: a geologic problem with an oceanographic solution. *Geology*, **6**, 708–712. https://pubs.geoscienceworld.org/gsa/geology/article-abstract/6/12/708/187537/The-terminal-Cretaceous-event-A-geologic-problem?redirectedFrom=fulltext

[13] D. A. Kring (2020) Understanding the K-T Boundary. Lunar and Planetary Institute. https://www.lpi.usra.edu/science/kring/Chicxulub/#:~:text=Although%20the%20K%2DT%20boundary%20has,K%2DPg%20mass%20extinction%20event (accessed 27 December 2023).

[14] American Museum of Natural History. Shelf Life 12: Six Extinctions in Six Minutes. https://www.amnh.org/shelf-life/six-extinctions (accessed 27 December 2023).

[15] What Derek Ager, in The Nature of the Stratigraphical Record (1973), actually said was "The history of any one part of the Earth, like the life of a soldier, consists of long periods of boredom and short periods of terror." https://todayinsci.com/A/Ager_Derek/AgerDerek-SoldierQuote800px.htm (accessed 27 December 2023).

[16] L. W. Alvarez, W. Alvarez, F. Asaro and H. V. Michel (1980) Extraterrestrial cause for the cretaceous-tertiary extinction. *Science*, **208**, 1095–1108. https://doi.org/10.1126/science.208.4448.1095

[17] D. A. Kring (2020) Discovering the Impact Site. Lunar and Planetary Institute Boundary. https://www.lpi.usra.edu/science/kring/Chicxulub/discovery/ (accessed 27 December 2023).

[18] P. Claeys and A. Morbidelli (2015) Late Heavy Bombardment. In: Gargaud, M., et al. (eds) *Encyclopedia of Astrobiology*. Springer, Berlin, Heidelberg. https://doi.org/10.1007/978-3-662-44185-5_869

[19] A. Mann (2018) Bashing holes in the tale of Earth's troubled youth. *Nature*, **553**, 393–395. https://doi.org/10.1038/d41586-018-01074-6
[20] K. Hansen and L. Dauphin (2018) Vredefort Crater. NASA Earth Observatory Website. https://earthobservatory.nasa.gov/images/92689/vredefort-crater (accessed 27 December 2023).
[21] Earth Impact Database, Sudbury, Planetary and Space Science Centre University of New Brunswick Fredericton, New Brunswick, Canada. https://www.passc.net/EarthImpactDatabase/New%20website_05-2018/Sudbury.html (accessed 27 December 2023).
[22] K. Hansen and J. Stevens (2020) Sudbury Impact Structure. NASA Earth Observatory website. https://earthobservatory.nasa.gov/images/148844/sudbury-impact-structure (accessed 27 December 2023).
[23] J. A. Petrus, D. E. Ames and B. S. Kamber (2015) On the track of the elusive Sudbury impact: geochemical evidence for a chondrite or comet bolide. *Terra Nova*, **27**, 9–20. https://onlinelibrary.wiley.com/doi/full/10.1111/ter.12125
[24] Kidd Creek Mine. Wikipedia. https://en.wikipedia.org/wiki/Kidd_Mine (accessed 27 December 2023).
[25] Shatter Cone. American Museum of Natural History. https://www.amnh.org/exhibitions/permanent/meteorites/meteorite-impacts/earth-impacts/all-craters-great-and-small/shatter-cone (accessed 27 December 2023).
[26] G. R. Osinski and L. Ferrière (2016) Shatter cones: (mis)understood? *Sciences Advances*, **2**. https://www.science.org/doi/10.1126/sciadv.1600616
[27] R. S. Dietz (1964) Sudbury structure as an astrobleme. *Journal of Geology*, **72**, 412–434. https://doi.org/10.1086/626999
[28] Planar Deformation Features. Wikipedia. https://en.wikipedia.org/wiki/Planar_deformation_features#:~:text=Planar%20deformation%20features%2C%20or%20PDFs,to%20the%20grain's%20crystal%20structure (accessed 27 December 2023).
[29] H. Ritchie and M. Roser (2014) "Natural Disasters". Published online at OurWorldInData.org. https://ourworldindata.org/natural-disasters#citation (accessed 27 December 2023).
[30] Science Daily (2017) New Study Analyzes Volcanic Fatalities in More Detail Than Ever Before. https://www.sciencedaily.com/releases/2017/10/171006101812.htm
[31] Skin Cancer. International Agency for Research on Cancer. World Health Organisation. https://www.iarc.who.int/cancer-type/skin-cancer/#:~:text=Skin%20cancers%20are%20the%20most,people%20died%20from%20the%20disease (accessed 27 December 2023).
[32] R. L. Holle (2016) The Number of Documented Global Lightning Fatalities. 24th International Lightning Detection Conference, San Diego. https://www.vaisala.com/sites/default/files/documents/Ron%20Holle.%20Number%20of%20Documented%20Global%20Lightning%20Fatalities.pdf (accessed 27 December 2023).
[33] L. Tondo (2021) Vesuvius Killed People of Pompeii in 15 Minutes, Study Suggests. The Guardian. https://www.theguardian.com/world/2021/mar/22/vesuvius-wiped-out-all-life-pompeii-15-minutes-study-pyroclastic-flow-cloud-gases-ash (accessed 27 December 2023).
[34] P. Jenniskens, O. P. Popova, D. O. Glazachev, E. D. Podobnaya and A. P. Kartashova (2019) Tunguska eyewitness accounts, injuries, and casualties. *Icarus*, **327**, 4–18. https://www.sciencedirect.com/science/article/abs/pii/S0019103518305104
[35] C. Trayner (1997) The Tunguska event. *Journal of the British Astronomical Association*, **107**, 117–130. https://articles.adsabs.harvard.edu/cgi-bin/nph-iarticle_query?bibcode=1997JBAA..107..117T&db_key=AST&page_ind=0&data_type=GIF&type=SCREEN_VIEW&classic=YES

[36] Protecting Planet Earth from Asteroids (2022) US Embassy and Consulates in Italy. https://it.usembassy.gov/protecting-planet-earth-from-asteroids/ (accessed 27 December 2023).

[37] John Anfinogenov, Larisa Budaeva, Dmitry Kuznetsov and Yana Anfinogenova (2014) John's Stone: A possible fragment of the 1908 Tunguska meteorite. *Icarus*, **243**, 139–147. https://www.sciencedirect.com/science/article/abs/pii/S0019103514004680

[38] A. Berard (2022) Traces of Ancient Ocean discovered on Mars. *Phys. Org* (retrieved 25 November 2022) https://phys.org/news/2022-10-ancient-ocean-mars.html (accessed 27 December 2023).

[39] The Meteoritical Bulletin Database. https://www.lpi.usra.edu/meteor/metbull.php (accessed 27 December 2023).

[40] Isotopes are varieties of an element all of which have the same number of protons but a different number of neutrons. Oxygen has three stable isotopes: oxygen-16 (^{16}O), oxygen-17 (^{17}O), and oxygen-18 (^{18}O). Materials from different sources in the Solar System have been found to have a different mix of these isotopes. Meteorites from Mars differ in their oxygen isotope composition compared to rocks from the Earth or Moon. So, analysing the oxygen isotope composition of a sample is a useful way of identifying whether it is Martian or not.

[41] H. Haack, R. C. Greenwood and H. Busemann (2016) Comment on "John's stone: a possible fragment of the 1908 Tunguska meteorite". (Anfinogenov et al. 2014, *Icarus*, **243**, 139–147) *Icarus*, **265**, 238–240. https://www.sciencedirect.com/science/article/abs/pii/S0019103515004236

[42] Y. Anfinogenova, J. Anfinogenov, L. Budaeva and D. Kuznetsov (2016) Response to the Comment by Haack et al. (2015) on the paper by Anfinogenov et al. (2014): John's stone: a possible fragment of the 1908 Tunguska meteorite. https://arxiv.org/ftp/arxiv/papers/1605/1605.01892.pdf

[43] The Mystery of the 2013 Chelyabinsk Superbolide Remains (2016) Science Blogging, Science 2.0 https://www.science20.com/news_articles/the_mystery_of_the_2013_chelyabinsk_superbolide_remains-165837

[44] Chelyabinsk Meteorite Fragment (2015) National Museum Scotland. https://www.nms.ac.uk/explore-our-collections/stories/natural-sciences/chelyabinsk-meteorite-fragment/ (accessed 27 December 2023).

[45] The Chelyabinsk Meteorite. Wikipedia. https://en.wikipedia.org/wiki/Chelyabinsk_meteor (accessed 27 December 2023).

[46] Meteoritical Bulletin Database Entry for Chelyabinsk. https://www.lpi.usra.edu/meteor/metbull.php?code=57165 (accessed 27 December 2023).

[47] NASA Explore, 5 Minute Read, Five Years after the Chelyabinsk Meteor: NASA Leads Efforts in Planetary Defense. https://www.nasa.gov/feature/five-years-after-the-chelyabinsk-meteor-nasa-leads-efforts-in-planetary-defense (accessed 27 December 2023).

[48] KPBS, Arts and Culture, NOVA Meteor Strike, Pioneer Productions/Channel 4. https://www.imdb.com/title/tt2787276/. https://www.kpbs.org/news/arts-culture/2013/03/15/nova-meteor-strike (accessed 27 December 2023).

[49] Meteoritical Bulletin Database Entry for Carancas. https://www.lpi.usra.edu/meteor/metbull.php?code=45817 (accessed 27 December 2023).

[50] J. Hecht (2007) Mysteries Remain over Peru Meteorite Impact. New Scientist Magazine. https://www.newscientist.com/article/dn12704-mysteries-remain-over-peru-meteorite-impact/ (accessed 27 December 2023).

[51] A. Boyle (2007) Sparks Fly over Meteorite. NBC News. https://www.nbcnews.com/sciencemain/sparks-fly-over-meteorite-6c10405428 (accessed 27 December 2023).

[52] G. Tancredi et al. (2009) A meteorite crater on Earth formed on September 15, 2007: the carancas hypervelocity impact. *Meteoritics and Planetary Science*, **44**, 1967–1984. https://onlinelibrary.wiley.com/doi/abs/10.1111/j.1945-5100.2009.tb02006.x#:~:text=Abstract%E2%80%94%20On%20September%2015%2C%202007,the%20site%20of%20the%20impact
[53] K. Yau, P. Weissman and D. Yeomans (1994) Meteorite falls in China and some related human casualty events. *Meteoritics*, **29**, 864–871. https://onlinelibrary.wiley.com/doi/abs/10.1111/j.1945-5100.1994.tb01101.x.
[54] O. Unsalan, A. Bayatali and P. Jenniskens (2020) Earliest evidence of a death and injury by a meteorite. *Meteoritics and Planetary Science*, **55**, 886–894. https://onlinelibrary.wiley.com/doi/10.1111/maps.13469.
[55] A. George (2019) In 1954, an Extraterrestrial Bruiser Shocked This Alabama Woman. Smithsonian Magazine. https://www.smithsonianmag.com/smithsonian-institution/1954-extraterrestrial-bruiser-shocked-alabama-woman-180973646/ (accessed 27 December 2023).
[56] DART: Double Asteroid Redirection Test. https://dart.jhuapl.edu/Mission/index.php (accessed 27 December 2023).

CHAPTER 5

[1] K. Hoekstra (2022) Plato's Myth: The Origins of the 'Lost' City of Atlantis. HistoryHit Website. https://www.historyhit.com/plato-and-the-lost-city-of-atlantis/ (accessed 28 December 2023).
[2] Atlantis of the Sands. Wikipedia. https://en.wikipedia.org/wiki/Atlantis_of_the_Sands (accessed 28 December 2023).
[3] R. Fiennes (1992) *Atlantis of the Sands: The Search for the Lost City of Ubar.* Published by Bloomsbury Pub Ltd, United Kingdom. ISBN 10: 0747513279/13: 9780747513278.
[4] S. Connor (1992) Lawrence's Lost "Atlantis" Found in Oman's Sands. Independent. https://www.independent.co.uk/news/uk/lawrence-s-lost-atlantis-found-in-oman-s-sands-1555493.html (accessed 28 December 2023).
[5] D. Millar (2017) Atlantis of the Sands and the Lost City of Ubar: Lost, Found, and Lost Again. Our Ancient History. https://www.phdeed.com/articles/atlantis-sands-and-lost-city-ubar-lost-found-and-lost-again (accessed 28 December 2023).
[6] M. Gladwell (2014) Trust No One. The New Yorker. https://www.newyorker.com/magazine/2014/07/28/philby (accessed 28 December 2023).
[7] John Banville (2014) A Spy among Friends: Kim Philby and the Great Betrayal – Review. The Guardian. https://www.theguardian.com/books/2014/mar/28/spy-among-friends-kim-philby-kgb-review (accessed 28 December 2023).
[8] Owen Bowcott (2005) Philby's Father Was Arrested. The Guardian newspaper. https://www.theguardian.com/politics/2005/nov/01/uk.past (accessed 28 December 2023).
[9] Arab News (2022) British Team to Retrace Steps of Epic Philby Trek across Saudi Arabia. https://www.arabnews.com/node/2171436/saudi-arabia (accessed 28 December 2023).
[10] M. Trotter (2020) The King's Man: Before Oil p Part 1. Aramco Expats. https://www.aramcoexpats.com/articles/the-king-s-man-before-oil-part-i/ (accessed 28 December 2023).
[11] M. Trotter (2020) The King's Man: Oil and Death - Part II. Aramco Expats. https://www.aramcoexpats.com/articles/the-king-s-man-oil-and-death-part-ii/ (accessed 28 December 2023).

[12] Z. Bilkadi (1986) The Wabar Meteorite Saudi. Aramco World. https://archive.aramcoworld.com/issue/198606/the.wabar.meteorite.htm (accessed 28 December 2023).
[13] The Prophet Hud. Wikipedia. https://en.wikipedia.org/wiki/Hud_(prophet) (accessed 28 December 2023).
[14] T. J. Abercrombie (1966) Saudi Arabia. Beyond the sands of Mecca. *National Geographic*, **1291**, 1–53. https://www.cuttersguide.com/pdf/National-Geographic/1966-01.pdf (accessed 28 December 2023).
[15] E. M. Shoemaker and J. C. Wynn (1997) Geology of the Wabar Meteorite Craters (Abstract #1660). 28th Lunar and Planetary Science Letters. https://www.lpi.usra.edu/meetings/lpsc97/pdf/1660.PDF
[16] J. Wynn and E. Shoemaker (1997) The Wabar Meteorite Impact Site, Ar-Rub' Al-Khali Desert, Saudi Arabia. https://volcanoes.usgs.gov/jwynn/3wabar.html (accessed 28 December 2023).
[17] E. Gnos et al. (2013) The Wabar impact craters, Saudi Arabia, revisited. *Meteoritics and Planetary Science*, **48**, 2000–2014. https://doi.org/10.1111/maps.12218
[18] J. R. Prescott, G.B. Robertson, C. Shoemaker, E.M. Shoemaker and J. Wynn (2004) Luminescence dating of the Wabar meteorite craters, Saudi Arabia). *Journal of Geophysical Research*, **109**, E01008. https://doi.org/10.1029/2003JE002136, https://agupubs.onlinelibrary.wiley.com/doi/full/10.1029/2003JE002136
[19] H. M. Basurah (2003) Estimating a new date for the Wabar meteorite impact. *Meteoritics and Planetary Science*, **38**, A155–A156. https://onlinelibrary.wiley.com/doi/abs/10.1111/j.1945-5100.2003.tb00324.x
[20] Meteoritical Bulletin Database Entry for the Wabar Iron Meteorite: https://www.lpi.usra.edu/meteor/metbull.php?code=24194 (accessed 28 December 2023).
[21] R. Hutchison (2004) *Meteorites A Petrologic, Chemical and Isotopic Synthesis.* Cambridge University Press. ISBN 139780521470100. https://assets.cambridge.org/97805214/70100/frontmatter/9780521470100_frontmatter.pdf
[22] E. R. D. Scott (2020) Iron Meteorites: Composition, Age, and Origin. Oxford Research Encyclopaedias, Planetary Science. https://oxfordre.com/planetaryscience/view/10.1093/acrefore/9780190647926.001.0001/acrefore-9780190647926-e-206 (accessed 28 December 2023).
[23] Q. Williams (1997) Why Is the Earth's Core So Hot? And How Do Scientists Measure Its Temperature? *Scientific American.* https://www.scientificamerican.com/article/why-is-the-earths-core-so/ (accessed 28 December 2023).
[24] T. H. Burbine, A. Meibom and R. P. Binzel (1996) Mantle material in the main belt: battered to bits? *Meteoritics and Planetary Science*, **31**, 607–620. https://doi.org/10.1111/j.1945-5100.1996.tb02033.x
[25] H. St. John Philby (1933) Rub' Al Khali: an account of exploration in the Great South Desert of Arabia under the auspices and patronage of His Majesty 'Abdul 'Aziz ibn Sa'ud, King of the Hejaz and Nejd and its dependencies. *The Geographical Journal*, **81**, 1–21. https://www.jstor.org/stable/1783888.

CHAPTER 6

[1] Dust Grains, Cosmos, Study Astronomy Online, Swinburne University, Melbourne, Australia. https://astronomy.swin.edu.au/cosmos/d/Dust+Grain (accessed 28/12/2023).
[2] NASA Explore. NASA's Webb Reveals Cosmic Cliffs, Glittering Landscape of Star Birth. https://www.nasa.gov/image-feature/goddard/2022/nasa-s-webb-reveals-cosmic-cliffs-glittering-landscape-of-star-birth (accessed 28/12/2023).

[3] J. Shelton (2018) A New Map for a Birthplace of Stars. Phys. Org. https://phys.org/news/2018-05-birthplace-stars.html (accessed 28/12/2023).
[4] A. Morbidelli, J. I. Lunine, D. P. O'Brien, S. N. Raymond and K. J. Walsh (2012) Building terrestrial planets. *Annual Review of Earth and Planetary Sciences*, **40**, 251–275.
[5] J. E. Chambers (2005) Planetary accretion in the inner solar system. *Earth and Planetary Science Letters*, **223**, 241–252. https://ui.adsabs.harvard.edu/abs/2004E&PSL.223..241C/abstract
[6] Artist's Impression of a Young Star Surrounded by a Protoplanetary Disc. European Southern Observatory https://www.eso.org/public/images/eso1436f/ (accessed 28/12/2023).
[7] M. Kaufman (2018) Planets Still Forming Detected in a Protoplanetary Disk. Astrobiology at NASA. https://astrobiology.nasa.gov/news/planets-still-forming-detected-in-a-protoplanetary-disk/ (accessed 28/12/2023).
[8] ALMA Image of the Protoplanetary Disc around HL Tauri. European Southern Observatory. https://www.eso.org/public/unitedkingdom/images/eso1436a/ (accessed 28/12/2023).
[9] A. Johansen and M. Lambrechts (2017) *Annual Review of Earth and Planetary Sciences*, **45**, 359–387. https://www.annualreviews.org/doi/full/10.1146/annurev-earth-063016-020226
[10] L. Grossman (2018) Jupiter May Be the Solar System's Oldest Planet. Science News Explores. https://www.snexplores.org/article/jupiter-may-be-solar-systems-oldest-planet (accessed 28/12/2023).
[11] Thomas S. Kruijer, Christoph Burkhardt, Gerrit Budde and Thorsten Kleine (2017) Age of Jupiter inferred from the distinct genetics and formation times of meteorites. *Proceedings of the National Academy of Sciences*, **114**, 6712–6716. https://www.pnas.org/doi/10.1073/pnas.1704461114.
[12] W. Clavin (2016) NASA Spitzer Telescope Investigating the Mystery of Migrating 'Hot Jupiters'. NASA Exoplanet Exploration. https://exoplanets.nasa.gov/news/1335/investigating-the-mystery-of-migrating-hot-jupiters/ (accessed 28/12/2023).
[13] On the Road Toward a Star, Planets Halt Their Migration (Artist Concept). NASA/JPL. https://www.jpl.nasa.gov/images/pia17242-on-the-road-toward-a-star-planets-halt-their-migration-artist-concept (accessed 28/12/2023).
[14] NASA Solar System Exploration. New Planetary Protection Board to Review Guidelines for Future Solar System and Beyond Exploration. https://solarsystem.nasa.gov/news/915/new-planetary-protection-board-to-review-guidelines-for-future-solar-system-and-beyond-exploration/ (accessed 28/12/2023).
[15] NASA Explore. Ceres Rotation and Occator Crater. False Colour Image of Dwarf Planet Ceres. https://solarsystem.nasa.gov/resources/846/ceres-rotation-and-occator-crater/?category=planets/dwarf-planets_ceres (accessed 28/12/2023).
[16] A. Nathues et al. (2020) Recent cryovolcanic activity at Occator crater on Ceres. *Nature Astronomy*, **4**, 794–801. https://www.nature.com/articles/s41550-020-1146-8
[17] Ceres, Earth & Moon Size Comparison. Wikimedia Commons https://commons.wikimedia.org/wiki/File:Ceres,_Earth_%26_Moon_size_comparison.jpg (accessed 28/12/2023).
[18] Kevin J. Walsh, A. Morbidelli, S. N. Raymond, D. P. O'Brian and A. M. Mandell (2012) Populating the asteroid belt from two parent source regions due to the migration of giant planets-"The Grand Tack". *Meteoritics and Planetary Science*, **47**, 1941–1947. https://onlinelibrary.wiley.com/doi/10.1111/j.1945-5100.2012.01418.x.
[19] NASA Explore. New Hubble Portrait of Mars. https://www.nasa.gov/feature/goddard/2016/new-hubble-portrait-of-mars (accessed 28/12/2023).
[20] Robin M. Canup and Erik Asphaug (2001) Origin of the Moon in a giant impact near the end of the Earth's formation. *Nature*, **412**, 708–712. https://www.nature.com/articles/35089010.

[21] J. A. Kegerreis et al. (2022) Immediate origin of the Moon as a post-impact satellite. *The Astrophysical Journal Letters*, **937**, L40. https://doi.org/10.3847/2041-8213/ac8d96.

[22] Frank Tavares (2022) NASA Explore. Collision May Have Formed the Moon in Mere Hours, Simulations Reveal. https://www.nasa.gov/feature/ames/lunar-origins- (accessed 28/12/2023).

[23] E. Asphaug and A. Reufer (2014) Mercury and other iron-rich planetary bodies as relics of inefficient accretion. *Nature Geoscience*, **7**, 564–568. https://www.nature.com/articles/ngeo2189

[24] P. Franco, A Izidoro, O.C. Winter, K.S. Torres and A. Amarante (2022) Explaining mercury via a single giant impact is highly unlikely. *Monthly Notices of the Royal Astronomical Society*, **515**, 5576–5586. https://doi.org/10.1093/mnras/stac2183.

[25] Michael K. Weisberg and Makoto Kimura (2012) The unequilibrated enstatite chondrites. *Chemie der Erde - Geochemistry*, **72**, 101–115. https://doi.org/10.1016/j.chemer.2012.04.003.

[26] NASA Explore. Planetary Smash-up. Artist's Concept Shows a Celestial Body about the Size of the Moon Slamming at Great Speed into a Body the Size of Mercury. https://www.nasa.gov/image-article/planetary-smash-up/ (accessed 28/12/2023).

[27] Wikimedia Commons. File:1850 Woman and Men in California Gold Rush.jpg. https://commons.wikimedia.org/wiki/File:1850_Woman_and_Men_in_California_Gold_Rush.jpg (accessed 28/12/2023).

[28] James M. D Day, Alan. D. Brandon and Richard J. Walker (2016) Highly siderophile elements in Earth, Mars, the Moon, and asteroids. *Reviews in Mineralogy & Geochemistry*, **81**, 161–238. https://doi.org/10.2138/rmg.2016.81.04

[29] A. H. Peslier, M. Schönbächler, H. Busemann and S.-I. Karato (2017) Water in the Earth's interior: Distribution and origin. *Space* Science Reviews, **212**, 743–810. https://link.springer.com/article/10.1007/s11214-017-0387-z

[30] Conel M. O'D. Alexander (2017) The origin of inner Solar System water. *Philosophical Transactions of the Royal Society*, **375**, 20150384. https://doi.org/10.1098/rsta.2015.0384

[31] David C. Rubie et al. (2011) Heterogeneous accretion, composition and core-mantle differentiation of the Earth. *Earth Planetary Science Letters*, **301**, 31–42. https://doi.org/10.1016/j.epsl.2010.11.030

[32] Richard C. Greenwood et al. (2023) Oxygen isotope evidence from Ryugu samples for early water delivery to Earth by CI chondrites. *Nature Astronomy*, **7**, 29-38. https://doi.org/10.1038/s41550-022-01824-7

[33] R. Black (2019) Black Fossil Site Reveals How Mammals Thrived after the Death of the Dinosaurs. Smithsonian Magazine. https://www.smithsonianmag.com/science-nature/fossil-site-reveals-how-mammals-thrived-after-death-of-dinosaurs-180973404/ (accessed 28/12/2023).

CHAPTER 7

[1] Here is the official Meteoritical Bulletin entry for the meteorite Git-Git which fell in Nigeria in 1947. Using the Meteoritical Bulletin Database, you can find all sorts of highly amusing names for meteorites. It is one of the fun things about studying space rocks. https://www.lpi.usra.edu/meteor/metbull.php?code=10919 (accessed 28/12/2023).

[2] Meteoritical Society Committee on Meteorite Nomenclature. Guidelines for Meteorite Nomenclature (February 1980 - Last Revised July 2015) https://www.lpi.usra.edu/meteor/docs/nc-guidelines.htm (accessed 28/12/2023).

[3] Llanfairpwllgwyngyllgogerychwyrndrobwllllantysiliogogogoch Is a Village on the Isle of Anglesey, Wales. https://www.llanfairpwllgwyngyllgogerychwyrndrobwllllantysiliogogogoch.co.uk/ (accessed 28/12/2023).
[4] Meteoritical Bulletin Database Entry for Camel Donga. https://www.lpi.usra.edu/meteor/metbull.php?code=5204 (accessed 28/12/2023).
[5] W. H. Cleverly, E. Jarosewich and B. Mason (1986) Camel Donga meteorite, a new eucrite from the Nullarbor plain, Western Australia. *Meteoritics*, **21**, 263–269. https://articles.adsabs.harvard.edu/full/1986Metic..21..263C
[6] Camel Donga 054 Official. Meteoritical Bulletin Database Entry https://www.lpi.usra.edu/meteor/metbull.php?code=73512 (accessed 28/12/2023).
[7] T. H. Burbine et al. (2001) Vesta, vestoids, and the howardite, eucrite, diogenite group: relationships and the origin of spectral differences. *Meteoritics* and Planetary Science, **36**, 761–781. https://articles.adsabs.harvard.edu/full/2001M%26PS...36..761B
[8] T. B. McCord, J. B. Adams and T. V. Johnson (1970) Asteroid Vesta: spectral reflectivity and compositional implications. *Science*, **168**, 1445–1447. https://www.jstor.org/stable/1730448?origin=ads
[9] H. P. Larson and U. Fink (1975) Infrared spectral observations of asteroid (4) Vesta. *Icarus*, **26**, 420–427. https://ui.adsabs.harvard.edu/abs/1975Icar...26..420L/abstract
[10] Thomas H. Burbine, Paul C. Buchanan, Tenzin Dolkar and Richard P. Binzel (2009) Pyroxene mineralogies of near-Earth vestoids. *Meteoritics and Planetary Science*, **44**, 1331–1341. https://onlinelibrary.wiley.com/doi/abs/10.1111/j.1945-5100.2009.tb01225.x
[11] Asteroid Main-Belt Distributions. NASA Jet Propulsion Laboratory. https://ssd.jpl.nasa.gov/diagrams/mb_hist.html (accessed 28/12/2023).
[12] R. P. Binzel and S. Xu (1993) Chips off of asteroid (4) Vesta: evidence for the parent body of basaltic achondrite meteorites. Science, 260, 186–191. https://ui.adsabs.harvard.edu/abs/1993Sci...260..186B/abstract
[13] NASA Explore Dawn. https://solarsystem.nasa.gov/missions/dawn/in-depth/#:~:text=During%20its%20nearly%20decade%2Dlong,processes%20that%20dominated%20its%20formation (accessed 28/12/2023).
[14] H. Y. McSween Jr. et al. (2013) Dawn; the Vesta-HED connection; and the geologic context for eucrites, diogenites, and howardites. *Meteoritics and Planetary Science*, **48**, 2090–2104. https://onlinelibrary.wiley.com/doi/10.1111/maps.12108.
[15] NASA Explore. Dawn Ion Propulsion System. https://solarsystem.nasa.gov/missions/dawn/technology/ion-propulsion/ (accessed 28/12/2023).
[16] NASA's Dawn Asteroid Mission Cancelled. Space.com. Published 03 March 2006. https://www.space.com/2116-nasas-dawn-asteroid-mission-cancelled.html (accessed November 2022).
[17] L. David (2006) Angry Scientists Confront NASA Officials. It was billed as an official NASA headquarters briefing to space scientists - but turned into a powder-keg of emotion. NBC News. https://www.nbcnews.com/id/wbna11829020 (accessed 28/12/2023).
[18] Dawn Mission Celebrates 10 Years in Space. Jet Propulsion Laboratory California Institute of Technology. https://www.jpl.nasa.gov/news/dawn-mission-celebrates-10-years-in-space (accessed 28/12/2023); The Veneneia Crater predates, and is partially obscured by, the Rheasilvia basin. https://en.wikipedia.org/wiki/Veneneia_(crater) (accessed 28/12/2023).
[19] H. Y. McSween Jr. et al. (2019) Differentiation and magmatic history of Vesta: constraints from HED meteorites and Dawn spacecraft data. *Chemie der Erde Geochemistry*, **79**, 125526. https://doi.org/10.1016/j.chemer.2019.07.008
[20] P. M. Schenk et al. (2022) A young age of formation of Rheasilvia basin on Vesta from floor deformation patterns and crater counts. *Meteoritics and Planetary Science*, **57**, 22–47. https://onlinelibrary.wiley.com/toc/19455100/2022/57/1

[21] A. M. Stickle, P. H. Schultz and D. A. Crawford (2015) Subsurface failure in spherical bodies: a formation scenario for linear troughs on Vesta's surface. *Icarus*, **247**, 18–34. https://www.sciencedirect.com/science/article/pii/S0019103514005302?via%3Dihub

[22] W. F. Bottke and M. Jutzi (2022) Collisional Evolution of the Main Belt as Recorded by Vesta. In: Marchi, S., Raymond, C., Russell, C. (eds) *Vesta and Ceres: Insights from the Dawn Mission for the Origin of the Solar System.* Cambridge Planetary Science, pp. 250–261. Cambridge: Cambridge University Press. https://doi.org/10.1017/9781108856324.020. https://www.cambridge.org/core/books/abs/vesta-and-ceres/collisional-evolution-of-the-main-belt-as-recorded-by-vesta/F49236BE81A129FB805DA0C9F1E68DA0

[23] T. B. McCord et al. (2012) Dark material on Vesta from the infall of carbonaceous volatile-rich material. *Nature*, **491**, 83–86. https://www.nature.com/articles/nature11561

CHAPTER 8

[1] U. B. Marvin (2007) Théodore Andre Monod and the lost Fer de Dieu meteorite of Chinguetti, Mauritania. *Geological Society, London, Special Publications*, **287**, 191–205. https://www.lyellcollection.org/doi/abs/10.1144/SP287.16

[2] Chinguetti, Mauritania. Wikipedia. https://en.wikipedia.org/wiki/Chinguetti (accessed 02/01/2024).

[3] Taken from U. B. Marvin (2007) Original source: T. Monod and B. Zanda (1992) Le Fer de Dieu: Histoire de la Meteorite de Chinguetti, Terres D'Aventure, Actes Sud, Arles. https://www.actes-sud.fr/node/15859

[4] K. C. Welten, P. A. Bland, S. S. Russell, M. M. Grady, M. W. Caffee, J. Masarik, A. J. T. Jull, H. W. Weber and L. Schultz (2001) Exposure age, terrestrial age and pre-atmospheric radius of the Chinguetti mesosiderite: not part of a much larger mass. *Meteoritics and Planetary Science*, **36**, 939–946. https://onlinelibrary.wiley.com/doi/abs/10.1111/j.1945-5100.2001.tb01931.x

[5] O. Eugster, G. F. Herzog, K. Marti and M. W. Caffee (2006) Irradiation Records, Cosmic-Ray Exposure Ages, and Transfer Times of Meteorites. In: Lauretta, D. S., McSween Jr., H. Y. (eds) *Meteorites and the Early Solar System* II. University of Arizona Press, Tucson, 943 pp., pp. 829–851. https://www.lpi.usra.edu/books/MESSII/9004.pdf

[6] N. Artemieva and E. Pierazzo (2009) The Canyon Diablo impact event: projectile motion through the atmosphere. *Meteoritics and Planetary Science*, **44**, 25–42. https://onlinelibrary.wiley.com/doi/abs/10.1111/j.1945-5100.2009.tb00715.x

[7] We can estimate the mass of the Fe de Dieu based on its dimensions. 40 m × 40 m × 100 m gives a volume of 160,000 m3 or 1.6×10^{11} cubic centimetres. The bulk density of mesosiderites is 4.25 g/cm3 as given by D. T. Britt and G. J. Consolmagno (2003) Stony meteorite porosities and densities: A review of the data through 2001. *Meteoritics and Planetary Science*, **38**, 1161–1180. That gives the Feu de Dieu a mass of $1.6 \times 10^{11} \times 4.25 = 6.8 \times 10^{11}$ g or 6.8×10^{8} kg or 680,000 metric tons.

[8] P. A. Bland and N. A. Artemieva (2006) The rate of small impacts on Earth. *Meteoritics and Planetary Science*, **41**, 607–631. https://onlinelibrary.wiley.com/doi/10.1111/j.1945-5100.2006.tb00485.x

[9] The lost Feu de Dieu meteorite remains a subject of popular interest. A documentary made with the participation of Professors Philip Bland and Sara Russell takes a look at the controversy surrounding this mysterious meteorite. The 100 Meter Meteorite That Just Disappeared | Fer de Dieu of Chinguetti | Spark. https://www.youtube.com/watch?v=YNDOAd0KBCU (accessed 02/01/2024).

[10] Meteoritical Bulletin Database, Official Classification of Chinguetti meteorite. https://www.lpi.usra.edu/meteor/metbull.php?code=5354 (accessed 02/01/2024).

[11] To understand what is going on here we need a little bit of background information. Atoms consist of a nucleus comprising protons and neutrons. They also have electrons, but we won't worry about that here. An element is defined by the number of protons its atoms contain. Change the number of protons and it becomes a new element. But the same element can have a variable number of neutrons. The different varieties of an element with different numbers of neutrons are called isotopes. As an example, stable oxygen atoms must always have 8 protons, but can have either 8, 9 or 10 neutrons. The number of protons and neutrons is added together so we say oxygen has three stable isotopes oxygen-16, oxygen-17 and oxygen-18.

Now here is the important bit for our current discussion. Not all isotopes of an element are stable and some may undergo radioactive decay. This is the case for one isotope of aluminium known as aluminium-26 which decays to stable magnesium-26. Unstable aluminium-26 doesn't hang around for long and is an example of a short-lived radioactive isotope. It was present in the very early Solar System but most of it had decayed away within a few million years. The decay process liberates a lot of energy and if an asteroid formed early enough it would have become totally molten as a result of this decay process.

For a more detailed treatment of short-lived isotopes see: Andrew M. Davis (2022) Short-lived nuclides in the early solar system: Abundances, origins, and applications. *Annual Review of Nuclear Particle Science*, **72**, 339–363. https://www.annualreviews.org/doi/full/10.1146/annurev-nucl-010722-074615

[12] P. J. Hevey and I. S. Sanders (2006) A model for planetesimal meltdown by 26Al and its implications for meteorite parent bodies. *Meteoritics and Planetary Science*, **41**, 95–106. https://journals.uair.arizona.edu/index.php/maps/article/viewFile/15227/15215

[13] The mineral olivine has the chemical composition $(Mg,Fe)_2SiO_4$ when of gem quality, it is known as peridot. https://en.wikipedia.org/wiki/Peridot. As a molten asteroid starts to solidify, it is one of the first minerals to form and because it is dense it sinks downwards whereas, plagioclase feldspar [14], which starts to crystalise later and is less dense, is more abundant at higher levels in the asteroid (Figure 8.5). Thus, a magma ocean will crystallise to form a lower olivine-rich zone and an upper plagioclase-rich zone. In the case of Vesta, the mineral orthopyroxene $(Mg,Fe)_2Si_2O_6$ takes the place of olivine but the same principals apply.

[14] Feldspars have a relatively complex chemical composition and vary between three end-members: Orthoclase $KAlSi_3O_8$; Albite $NaAlSi_3O_8$; and Anorthite $CaAl_2Si_2O_8$ compared to olivine, feldspar has a lower density and so is much more abundant in the outer part of an asteroid that went through a molten phase.

Feldspar. Imerys Group website https://www.imerys.com/minerals/feldspar#:~:text=Feldspar%20is%20the%20name%20given,about%2050%25%20of%20all%20rocks (accessed 02/01/2024).

[15] E. Asphaug (2017) Signatures of Hit and Run Collisions. In: Elkins-Tanton, L. T., Weiss, B. P. (eds) Planetesimals: Early Differentiation and Consequences for Planets. Cambridge University Press. ISBN 978-1-107-11848-5. https://arxiv.org/ftp/arxiv/papers/1810/1810.05797.pdf

[16] M. K. Haba, Jorn-Frederik Wotzlaw, Yi-Jen Lai, Akira Yamaguchi and Maria Schonbachler (2019) Mesosiderite formation on asteroid (4) Vesta by a hit-and-run collision. *Nature Geoscience*, **12**, 510–515. https://www.nature.com/articles/s41561-019-0377-8#:~:text=Mesosiderites%20are%20stony%2Diron%20meteorites,break%2Dup%20of%20differentiated%20asteroids

[17] R. C. Greenwood, T. H. Burbine, M. F. Miller and I. A. Franchi (2017) Melting and differentiation of early-formed asteroids: the perspective from high precision oxygen isotope studies. *Chemie der Erde – Geochemistry*, **77**(1), 1–43. https://www.sciencedirect.com/science/article/pii/S0009281916301994?via%3Dihub

[18] K. Sugiura, Makiko K. Haba and Hidenori Genda (2022) Giant impact onto a Vesta-like asteroid and formation of mesosiderites through mixing of metallic core and surface crust. *Icarus*, **379**, 114949. https://doi.org/10.1016/j.icarus.2022.114949

[19] E. Palomba et al. (2013) Mesosiderites on Vesta: A Hyper-Spectral VIS-NIR Investigation. 44th Lunar and Planetary Science Conference. Abstract # 2245. https://www.lpi.usra.edu/meetings/lpsc2013/pdf/2245.pdf

[20] S. Iannini LeLarge et al. (2022) Asteroids, accretion, differentiation and break-up in the Vesta source region. Evidence from cosmochemistry of mesosiderites. *Geochimica et Cosmochimica Acta*, **329**, 135–151. https://www.sciencedirect.com/science/article/pii/S0016703722002186

CHAPTER 9

[1] Meteorites enter the upper atmosphere at speeds from about 11 km per second (25,000 mph) up to about 72 km/second (160,000 mph). Smaller meteorites will be slowed by the Earth's atmosphere and reach "terminal velocity", which for most meteorites will be about 200 to 400 mph. On impact with the Earth's surface, all they will do is put a small dent in the ground. But the "Big Ones" will retain a significant component of their initial velocity and hence slam into the surface of the Earth creating an impact crater. So, what sort of size are the "Big Ones"? Well, meteorites above 9,000 kg (9 metric tons) retain about 6% of their cosmic velocity. Assuming it is a stony type (bulk density about 3 grams per cubic centimetre) with a spherical geometry, a 9,000 kg meteorite has a diameter of about 2 m. A stony meteorite weighing 900,000 kg (900 metric tons, 8.5 m diameter), retains about 70% of its cosmic velocity. And finally, a 90,000,000 kg stony meteorite (90,000 metric tons, about 40 m diameter) will lose almost none of its cosmic velocity. (Source: American Meteor Society. Fireball FAQs) https://www.amsmeteors.org/fireballs/faqf/#8; Artemieva and Pierazzo (2009) estimate that the Canyon Diablo meteorite which formed Meteor Crater had a mass prior to entering the Earth's atmosphere of between 400,000,000 kg (400,000 metric tons) and 1,200,000,000 kg (1.2 million metric tons). That represents a lump of metal (Canyon Diablo is an iron meteorite) between 46 and 66 m* in diameter, which is more or less the height of a 15 storey office building! But the Canyon Diablo meteorite didn't make it through the atmosphere unscathed. Artemieva and Pierazzo (2009) estimate that it lost between 30% and 70% of its original mass during atmospheric entry and also underwent some fragmentation. Based on their lower mass estimate and higher rate of atmospheric attrition, that would still make the impacting mass about 120,000 metric tons. In terms of its speed, they suggest that it may have entered the atmosphere travelling at about 18 km/s and that it hit the ground at no less than 15 km/s, which is 33,554 mph, or New York to London in 6.2 minutes! So, slightly faster than Concorde used to do it in! Well, a lot faster actually! And you can see why it might have made a bit of a dent in the Arizona landscape on arrival. The result of the impact would have been that it was brought to an instantaneous halt resulting in a huge amount of kinetic energy being deposited into the rocks of the Arizona desert, with cataclysmic results (see main text for further details).

NataliaArtemieva and Elisabetta Pierazzo (2009) The Canyon Diablo impact event: projectile motion through the atmosphere. *Meteoritics and Planetary Science*, **44**, 25–42. https://onlinelibrary.wiley.com/doi/10.1111/j.1945-5100.2009.tb00715.x.

* It is important to remember that iron meteorites are much denser than stony types having a density of approximately 7.5 g/cm3 compared to about 3 g/cm3 in the case of stones.

G. J. Consolmagno and D. T. Britt (1998) The density and porosity of meteorites from the Vatican collection. *Meteoritics and Planetary Science*, **33**, 1231–1241. https://doi.org/10.1111/j.1945-5100.1998.tb01308.x.

[2] Meteor Crater has been known by other names in the past including: Coon Mountain, Coon Butte, Crater Mountain, and Meteor Mountain. In the scientific literature, it is often known as Barringer Crater after Daniel Moreau Barringer (1860–1929), who championed the proposition that it was formed by a meteorite impact.

[3] Winslow, Arizona is also famous for being the town sung about in the Eargles' song Take It Easy. https://www.songfacts.com/lyrics/eagles/take-it-easy (accessed 02/01/2024).

[4] Kathryn Hansen (2021) Arizona's Meteor Crater. NASA's Earth Observatory. https://earthobservatory.nasa.gov/images/148384/arizonas-meteor-crater#:~:text=Meteor%20Crater%20measures%200.75%20miles,million%20metric%20tons%20of%20rock (accessed 02/01/2024).

[5] D. A. Kring (1997) Air blast produced by the Meteor Crater impact event and a reconstruction of the affected environment. *Meteoritics and Planetary Science*, **32**, 517–530. https://onlinelibrary.wiley.com/doi/epdf/10.1111/j.1945-5100.1997.tb01297.x.

[6] The Mineralogical Record. Albert Edward Foote (1846–1995) https://mineralogicalrecord.com/new_biobibliography/foote-albert-edward/ (accessed 02/01/2024).

[7] Wikipedia entry for Albert E. Foote. https://en.wikipedia.org/wiki/Albert_E._Foote (accessed 02/01/2024).

[8] H. H. Nininger (1951) A résumé of researches at the Arizona Meteorite Crater. *The Scientific Monthly*, **72**(2), 75–86. American Association for the Advancement of Science. https://www.jstor.org/stable/20303.

[9] D. A. Kring (2017) *Guidebook to the Geology of Barringer Meteorite Crater (Arizona a.k.a. Meteor Crater) 2nd Edition*. Lunar and Planetary Institute, Contribution Number 2040. https://www.lpi.usra.edu/publications/books/barringer_crater_guidebook/.

[10] A. E. Foote (1891) A new locality for meteoric iron with a preliminary notice of the discovery of diamond in the iron. *American Journal of Science*, **s3-42**(251), 413–417. https://doi.org/10.2475/ajs.s3-42.251.413.

[11] Widmanstätten pattern is an intergrowth of Ni-rich and Ni-poor mineral phases found in iron meteorites and some stony-iron meteorites. It forms in the solid state during very slow cooling. Two iron-rich phases commonly found in iron meteorites are nickel-rich taenite, and nickel-poor kamacite. These essentially unmix from a high-temperature homogeneous state along distinct orientations. The result is a distinctive crosshatch pattern. A nice example of Widmanstätten pattern in the Canyon Diablo meteorite can be seen on the Natural History Museum blog: Kerry Lotzof. Types of meteorites, The Natural History Museum. https://www.nhm.ac.uk/discover/types-of-meteorites.html (accessed 02/01/2024).

[12] Official Meteoritical Bulletin Entry for Canyon Diablo. https://www.lpi.usra.edu/meteor/metbull.php?code=5257 (accessed 02/01/2024).

[13] Canyon Diablo is an iron meteorite of the IAB group. The IABs are an interesting group because they do not appear to have formed in the core of an asteroid in the way that many iron meteorites are likely to have done. These are the so-called "magmatic irons" because they probably crystallised from a totally molten state. The idea is that when as asteroid melts liquid iron, due to its high density, ponds in the centre of the asteroid forming a dense core. In contrast, groups like the IABs are known as "non-magmatic irons". IAB irons like Canyon Diablo show chemical and isotopic similarities to a group of stony meteorites known as the winonaites. Both groups are thought to have come from a single asteroid that was catastrophically destroyed before it could fully separate into a core, mantle, and crust.

G. K. Benedix, T.J. McCoy, K. Keil and S.G. Love (2000) A petrologic study of the IAB iron meteorites: constraints on the formation of the IAB-Winonaite parent body. *Meteoritics and Planetary Science*, **35**, 1127–1141. https://onlinelibrary.wiley.com/doi/10.1111/j.1945-5100.2000.tb01502.x

[14] G. K. Gilbert (1896) The origin of hypotheses, illustrated by the discussion of a topographic problem. *Science*, **3**(53), 1–13. https://www.science.org/doi/10.1126/science.3.53.1

[15] G. K. Gilbert (1893) The moon's face, a study of the origin of its features. Address as retiring president delivered December 10, 1892. *Philosophical Society of Washington Bulletin*, **12**, 241–292. https://babel.hathitrust.org/cgi/pt?id=hvd.32044080587637&view=1up&seq=5

[16] B. Barringer (1964) Daniel Moreau Barringer (1860–1929) and his Crater (the beginning of the Crater Branch of Meteoritics). *Meteoritics*, **2**, 183–200. https://adsabs.harvard.edu/full/1964Metic...2..183B.

[17] The Barringer Crater Company. The Crater - Fascinating Science and Unique History. https://barringercrater.com/the-crater (accessed 02/01/2024)

[18] D. M. Barringer (1905) Coon Mountain and its crater. *Proceedings of the Academy of Natural Sciences of Philadelphia*, **57**, 861–886. https://www.lpi.usra.edu/publications/books/barringer_crater_guidebook/BarringerReports/Barringer_CoonMountainAndItsCrater_1905.pdf

[19] Shaping the Planets: Impact Cratering. LPI Learning. https://www.lpi.usra.edu/education/explore/shaping_the_planets/impact-cratering/ (accessed 02/01/2024).

[20] H. Plotkin and R. S. Clarke Jr. (2008) Harvey Nininger's 1948 attempt to nationalize meteor crater. *Meteoritics and Planetary Science*, **43**, 1741–1756. https://doi.org/10.1111/j.1945-5100.2008.tb00640.x.

[21] Some sources suggest that the American Meteorite Museum opened in 1942 rather than 1946 e.g. Mary-Elizabeth Zucolotto and Amanda Tosi (2020) Seeking the Nininger's Museums Meteorite Times Magazine. https://www.meteorite-times.com/seeking-the-niningers-museums-from-the-ruins-near-the-arizona-crater-to-the-unknown-building-that-now-belongs-to-a-hotel/ (accessed 02/01/2024).

[22] E. C. T. Chao, E. M. Shoemaker and B. M. Madsen (1960) First natural occurrence of coesite. *Science*, **132**, 220–222. https://www.jstor.org/stable/1705158?origin=ads.

[23] Coesite is a variety of SiO2 that is formed at ultra-high pressures. Until its discovery at Meteor Crater in 1960, it had only been known from synthetic examples. Minerals.net. The Mineral & Gemstone Kingdom. https://www.minerals.net/mineral/coesite.aspx (accessed 02/01/2024).

[24] E. M. Shoemaker (1963) Impact Mechanics at Meteor Crater, Arizona. In: Kuiper, G. P., Middlehurts, B. (eds) *The Moon Meteorites and Comets*. The University of Chicago Press, Chicago, p. 301. https://articles.adsabs.harvard.edu/pdf/1963mmc..book..301S.

[25] The Teapot Ess test took place in March 1955 and involved an underground explosion equivalent to 1.2 kilotons of TNT. The footage of the explosion provides a visual impression of what the aftermath of the Meteor Crater impact event might have looked like. It also shows how few precautions were taken to protect the spectators from the effects of the nuclear fallout.
US Nuclear Tests, Info Gallery, Radiochemical Society. https://www.radiochemistry.org/history/nuke_tests/teapot/index.html
Footage of Teapot Ess Nuclear Detonation: Atom Central, YouTube. https://www.youtube.com/watch?v=I9ahoAMAGL8 (accessed 02/01/2024).

[26] There is a significant level of uncertainty concerning the amount of energy released during the impact event that formed meteor crater. However, as discussed by David Kring in his authoritative Meteor Crater guidebook (pp. 119–1120), a value close to 10 million tons of TNT appears to be a reasonable estimate. D. A. Kring (2017) *Guidebook to the Geology of Barringer Meteorite Crater (Arizona a.k.a. Meteor Crater) 2nd Edition.* Lunar and Planetary Institute, Contribution Number 2040. https://www.lpi.usra.edu/publications/books/barringer_crater_guidebook/

[27] Sedan Crater. USGS Earth Resources Observation and Science Center (EROS). https://eros.usgs.gov/media-gallery/earthshot/sedan-crater (accessed 02/01/2024).

[28] W. S. Kiefer (2003) *Impact Craters in the Solar System Space Science Reference Guide, 2nd Edition.* Lunar and Planetary Institute. https://www.lpi.usra.edu/science/kiefer/Education/SSRG2-Craters/craterstructure.html.

[29] With increasing size, impact craters throughout the Solar System show a change in morphology, from “simple” craters such as Meteor Crater to “complex” craters such as Tycho on the Moon (Figure 9.9). The transition diameter from simple to complex craters varies from one Solar System body to another and depends on the gravity of the body and the strength of the rocks in which the crater is formed. On Earth the transition diameter is about 3 km, it is about 20 km on the Moon and 7 km on Mars. Further information on crater morphology can be found here: Center for lunar science and exploration, Impact Cratering Lab, Lunar and Planetary Institute. https://www.lpi.usra.edu/exploration/education/hsResearch/crateringLab/lab/part1/background/#:~:text=IMPACT%20CRATERING%20MORPHOLOGY,material%20ejected%20from%20the%20crater (accessed 02/01/2024).

[30] Tycho Crater’s Central Peak on the Moon. NASA Solar System Exploration. https://solarsystem.nasa.gov/resources/2265/tycho-craters-central-peak-on-the-moon/ (accessed 02/01/2024).

[31] S. Schwenzer (2016) Study Sheds Light on Violent Asteroid Crash That Caused Mysterious ‘Crater Rings’ on the Moon. The Conversation. https://theconversation.com/study-sheds-light-on-violent-asteroid-crash-that-caused-mysterious-crater-rings-on-the-moon-68093 (accessed 02/01/2024).

[32] Apollo Field Operations Test III, Meteor Crater, AZ, May 18–20, 1965. Astropedia, USGS Astrogeology Science Center. https://astrogeology.usgs.gov/search/map/RPIF/Videos/apollofieldoperationsmeteorcrater?p=6&pb=1 (accessed 02/01/2024).

CHAPTER 10

[1] E. Osterloff, Wold Cottage: The Stone That Proved Meteorites Come from Space. Natural History Museum, London. https://www.nhm.ac.uk/discover/the-wold-cottage-meteorite.html (accessed 02/01/2024).

[2] C. T. Pillinger and J. M. Pillinger (1996) The Wold Cottage meteorite: not just any ordinary chondrite. *Meteoritics and Planetary Science*, **31**, 589–605. https://onlinelibrary.wiley.com/doi/abs/10.1111/j.1945-5100.1996.tb02032.x
[3] Eric Hutton (2007) Stretchleigh Meteorite, UK and Ireland meteorite page. https://www.meteoritehistory.info/UKIRELAND/C17.HTM (accessed 02/01/2024).
[4] A. J. King et al. (2022) The Winchcombe Meteorite, a unique and pristine witness from the outer solar system. *Sciences Advances*, **8**. https://www.science.org/doi/10.1126/sciadv.abq3925
[5] S. S. Russell et al. (2023) Recovery and Curation of the Winchcombe (CM2) meteorite. *Meteoritics and Planetary Science.* https://doi.org/10.1111/maps.13956
[6] Sarah McCullen et al. (2023) The Winchcombe fireball - that lucky survivor. *Meteoritics and Planetary Science.* https://doi.org/10.1111/maps.13977
[7] Queenie H. S. Chan et al. (2023) The amino acid and polycyclic aromatic hydrocarbon compositions of the promptly recovered CM2 Winchcombe carbonaceous chondrite organic material in the Winchcombe meteorite. *Meteoritics and Planetary Science.* https://doi.org/10.1111/maps.13936
[8] C. M. O'D. Alexander et al. (2017) The nature, origin and modification of insoluble organic matter in chondrites, the possibly interstellar source of Earth's C and N. *Chemie der Erde – Geochemistry*, **77**, 227–256. https://www.sciencedirect.com/science/article/pii/S0009281916301350?via%3Dihub
[9] D. P. Glavin et al. (2018) The Origin and Evolution of Organic Matter in Carbonaceous Chondrites and Links to Their Parent Bodies. In: Abreu, N. M. (ed) *Primitive Meteorites and Asteroids. Physical, Chemical and Spectroscopic Observations Paving the Way to Exploration 2018*, pp. 205–271. https://doi.org/10.1016/B978-0-12-813325-5.00003-3. https://science.gsfc.nasa.gov/sed/content/uploadFiles/publication_files/Chapter-3---The-Origin-and-Evolution-of-Organic-Matter_2018_Primitive-Meteor.pdf
[10] K. J. Walsh, A. Morbidelli, S.N. Raymond, D.P.O'Brien and A.M. Mandell (2012) Populating the asteroid belt from two parent source regions due to the migration of giant planets-"The Grand Tack". *Meteoritics and Planetary Science*, **47**, 1941–1947. https://onlinelibrary.wiley.com/doi/10.1111/j.1945-5100.2012.01418.x
[11] Martin D. Suttle et al. (2022) The Winchcombe meteorite - a regolith breccia from a rubble-pile CM chondrite asteroid. *Meteoritics and Planetary Science.* https://doi.org/10.1111/maps.13938
[12] Richard C. Greenwood et al. (2023) The formation and aqueous alteration of CM2 chondrites and their relationship to CO3 chondrites: a fresh isotopic (O, Cd, Cr, Si, Te, Ti and Zn) perspective from the Winchcombe CM2 fall. *Meteoritics and Planetary Science.* https://onlinelibrary.wiley.com/doi/10.1111/maps.13968
[13] Y. Amelin et al. (2002) Lead isotopic ages of chondrules and calcium-aluminum-rich inclusions. Science, 297, 1678–1683. https://www.science.org/doi/abs/10.1126/science.1073950.
[14] The element oxygen actually comes in three distinct flavours. The more scientific word is isotopes. Isotopes are different varieties of an element all of which have the same number of protons, but have a different number of neutrons. The three isotopes of oxygen are oxygen-16, oxygen-17, and oxygen-18. All have eight protons, but oxygen-16 has eight neutrons, oxygen-17 has nine neutrons, and oxygen-18 has ten neutrons. The abundance of these isotopes is measured using an instrument called a mass spectrometer. It turns out that oxygen isotopes can be used to help figure out what group a meteorite might belong to.

CHAPTER 11

[1] Ellen Castelow. The Battle of Stamford Bridge. Historic UK. https://www.historic-uk.com/HistoryMagazine/DestinationsUK/The-Battle-of-Stamford-Bridge/(accessed 02/01/2024).

[2] What Happened at the Battle of Hastings? English Heritage. https://www.english-heritage.org.uk/visit/places/1066-battle-of-hastings-abbey-and-battlefield/history-and-stories/what-happened-battle-hastings/(accessed 02/01/2024).
[3] Apollo 8, Smithsonian. National Air and Space Museum. https://airandspace.si.edu/explore/stories/apollo-missions/apollo-8 (accessed 02/01/2024).
[4] Roy S. Clarke Jr. et al. (1971) The Allende, Mexico, meteorite shower. *Smithsonian Contributions to the Earth Sciences*, **5**, 1–53. https://repository.si.edu/handle/10088/809
[5] T. J. McCoy and C. M. Corrigan (2021) The Allende meteorite: landmark and cautionary tale. *Meteoritics and Planetary Science*, **56**, 5–7. https://onlinelibrary.wiley.com/doi/epdf/10.1111/maps.13613
[6] Allende meteorite. Wikipedia. https://en.wikipedia.org/wiki/Allende_meteorite.
[7] E. A. King Jr., E. Schonfield, K. A. Richardson and J. S. Eldridge (1969) Meteorite fall at Peublito de Allende, Chihuahua, Mexico: preliminary information. *Science*, **163**, 928–929. https://www.science.org/doi/10.1126/science.163.3870.928
[8] Meteoritical Bulletin Database Official Entry for the Allende Meteorite. https://www.lpi.usra.edu/meteor/metbull.php?code=2278 (accessed 02/01/2024).
[9] M. C. Michel-Levy (1968) Un chondre exceptionnel dans la meteorite de Vigarano. *Bulletin de la Société Française de Minéralogie et de Cristallographie*, **91**, 212–214. https://www.persee.fr/doc/bulmi_0037-9328_1968_num_91_2_6215
[10] L. Grossman (1972) Condensation in the primitive solar nebula. *Geochimica et Cosmochimica Acta*, **36**, 597–619. https://geosci.uchicago.edu/~grossman/G72GCA.pdf
[11] Y. Amelin, A. N. Krot, I. D. Hutcheon and A. A. Ulyanov (2002) Lead isotopic ages of chondrules and calcium-aluminum-rich inclusions. *Science*, **297**, 1678–1683. https://www.jstor.org/stable/3832321
[12] Alexander N. Krot (2002) Dating the Earliest Solids in Our Solar System. PSRD Discoveries. https://www.psrd.hawaii.edu/Sept02/isotopicAges.html (accessed 02/01/2024).
[13] Dark Nebulae and Star Formation in Taurus. Astronomy Picture of the Day 21 March 2023. https://apod.nasa.gov/apod/ap230321.html (accessed 02/01/2024).
[14] Gregory A. Brennecka et al. (2020) Astronomical context of Solar System formation from molybdenum isotopes in meteorite inclusions. *Science*, **370**, 837–840. https://www.science.org/doi/10.1126/science.aaz8482
[15] David A. J. Seargent (1990) The Murchison meteorite: circumstances of its fall. *Meteoritics*, **25**, 341–342. https://articles.adsabs.harvard.edu/cgi-bin/nph-iarticle_query?1990Metic..25..341S&defaultprint=YES&page_ind=0&filetype=.pdf
[16] M. Zolensky, S. Russell and A. Brearley (2021) The fall of the Murchison Meteorite. *Meteoritics and Planetary Science*, **56**, 8–10. https://onlinelibrary.wiley.com/doi/10.1111/maps.13596
[17] Meteoritical Bulletin Database Official Entry for the Murchison meteorite. https://www.lpi.usra.edu/meteor/metbull.php?code=16875 (accessed 02/01/2024).
[18] F. Pepper (2019) When a Space Visitor Came to Country Victoria. ABC Science. https://www.abc.net.au/news/science/2019-10-02/murchison-meteorite-50th-anniversary-1969-science-geology/11528644 (accessed 02/01/2024).
[19] M. D. Suttle, A. J. King, P. F. Schofield, H. Bates and S. S. Russel (2021) The aqueous alteration of CM chondrites, a review. *Geochimica et Cosmochimica Acta*, **299**, 219–256. https://www.sciencedirect.com/science/article/pii/S0016703721000363
[20] E. Zinner (2002) Using Aluminum-26 as a Clock for Early Solar System Events. PSRD Discoveries. https://www.psrd.hawaii.edu/Sept02/Al26clock.html (accessed 02/01/2024).
[21] R. Findlay (2022) Can a Space Rock from Costa Rica Reveal the Origin of Water on Earth? Open Learn. https://openlearn.medium.com/can-a-space-rock-from-costa-rica-reveal-the-origin-of-water-on-earth-2fd648526b15 (accessed 02/01/2024).

[22] U. Smith. The Murchison Meteorite. Museums Victoria. https://museumsvictoria.com.au/article/the-murchison-meteorite/ (accessed 02/01/2024).
[23] S. Pizzarello and E. Shock (2010) The organic composition of carbonaceous meteorites: the evolutionary story ahead of biochemistry. *Cold Spring Harbor Perspectives in Biology*, **201**, a002105. https://www.ncbi.nlm.nih.gov/pmc/articles/PMC2829962/
[24] J. Matson (2010) Meteorite That Fell in 1969 Still Revealing Secrets of the Early Solar System. Scientific American. https://www.scientificamerican.com/article/murchison-meteorite/ (accessed 02/01/2024).
[25] M. A. Sephton (2002) Organic compounds in carbonaceous chondrites. Natural Products Report, **19**, 292–311. https://www.researchgate.net/publication/11245192_Organic_Compounds_in_Carbonaceous_Meteorites
[26] D. P. Glavin et al. (2021) Extraterrestrial amino acids and L-enantiomeric excesses in the CM2 carbonaceous chondrites Aguas Zarcas and Murchison. *Meteoritics and Planetary Science*, **56**, 148–173. https://onlinelibrary.wiley.com/doi/10.1111/maps.13451

CHAPTER 12

[1] ANSMET, The Antarctic Search for Meteorites, 2023/2024 Field Season, Case Western Reserve University. https://caslabs.case.edu/ansmet/2023/ (accessed 02/01/2024).
[2] M. Yoshida (2010) Discovery of the Yamato Meteorites in 1969. *Polar Sci.*, **3**, 272–284. https://www.sciencedirect.com/science/article/pii/S1873965209000589
[3] Official Meteoritical Bulletin database entry for the Adelie Land Meteorite. https://www.lpi.usra.edu/meteor/metbull.php?code=378 (accessed 02/01/2024).
[4] Ross Pogson (2023) The Centenary of the Adelie Land Meteorite. Australian Museum. https://australian.museum/blog-archive/science/centenary-adelie-meteorite/ (accessed 02/01/2024).
[5] D. Sears (22 March 1979) Rocks on the Ice. New Scientist. https://dsears.hosted.uark.edu//publications/pub%20by%20year/1979%20papers/sears%201979d.pdf (accessed 02/01/2024).
[6] Ralph P. Harvey, John W. Schutt and Christian Koeberl (2020) In Memoriam William A. Cassidy, The Meteoritical Society. https://meteoritical.org/news/william-cassidy-1928-2020 (accessed 02/01/2024).
[7] W. A. Cassidy (2012) *Meteorites, Ice, and Antarctica*. Cambridge University Press. https://www.cambridge.org/gb/academic/subjects/physics/planetary-systems-and-astrobiology/meteorites-ice-and-antarctica-personal-account?format=PB&isbn=9781107403918
[8] Ralph Harvey (2019) 50 Years Ago Today - The Discovery of the First Yamato Meteorite. ANSMET, The Antarctic Search for Meteorites, Case Western Reserve University. https://caslabs.case.edu/ansmet/2019/12/22/50-years-ago-today/ (accessed 02/01/2024). See also: Ralph Harvey (2003) The origin and significance of Antarctic meteorites. *Chemie der Erde – Geochemistry*, **63**, 93–147. https://artscimedia.case.edu/wp-content/uploads/sites/111/2016/09/14205314/Origin-and-significance.pdf
[9] ANSMET - The Antarctic Search for Meteorites, Case Western Reserve University. https://caslabs.case.edu/ansmet/faqs/ and Kevin Righter (NASA, Johnson Space Center) personal communication (accessed 02/01/2024).
[10] Akira Yamaguchi et al. (2019) Japanese Antarctic meteorite collection: past, current, and future. 82nd Annual Meeting of The Meteoritical Society abstract #6224 https://www.hou.usra.edu/meetings/metsoc2019/pdf/6224.pdf
[11] G. W. Evatt et al (2020) The spatial flux of Earth's meteorite falls found via Antarctic data. *Geology*, **48**, 683–683. https://doi.org/10.1130/G46733.1

[12] Veronica Tollenaar et al. (2022) Unexplored Antarctic meteorite collection sites revealed through machine learning. *Science Advances* **8**. https://doi.org/10.1126/sciadv.abj8138
[13] Hiroshi Naraoka (2019) In Memoriam Keizo Yanai. The Meteoritical Society. https://meteoritical.org/news/keizo-yanai-1941-2018 (accessed 02/01/2024).
[14] FAQs, How Does ANSMET Search for Meteorites? ANSMET, The Antarctic Search for Meteorites, Case Western Reserve University. https://caslabs.case.edu/ansmet/faqs/ (accessed 02/01/2024).
[15] NASA Curation of Antarctic Meteorites. Initial Characterization of Samples. https://curator.jsc.nasa.gov/antmet/collection_curation.cfm?section=characterization (accessed 02/01/2024).
[16] A. Eisenstadt (2022) Get to Know the Geologist Collecting Antarctic Meteorites. Smithsonian Voices. https://www.smithsonianmag.com/blogs/national-museum-of-natural-history/2022/01/11/get-to-know-the-geologist-collecting-antarctic-meteorites/ (accessed 02/01/2024).
[17] NASA Curation, Antarctic Meteorites. Antarctic Meteorite Newsletter. https://curator.jsc.nasa.gov/antmet/amn/amn.cfm#n462 (accessed 02/01/2024).
[18] NASA Curation, Antarctic Meteorites, Sample Requests, Allocations and Loans. https://curator.jsc.nasa.gov/antmet/requests.cfm?section=general (accessed 02/01/2024).
[19] Belgian-Dutch Team Map Meteorites in Antarctica (2022). The Brussels Times. https://www.brusselstimes.com/203669/belgian-dutch-team-map-meteorites-in-antarctica (accessed 02/01/2024).
[20] Z. P. Xia, B. K. Miao, J. Zhang, K. Y. Zhao and Y. L. Sun (2017) Meteorite collection by CHINARE in Antarctica. *Journal of Global Change Data & Discovery*, **1**, 331–335.
[21] Luigi Folco (2018) Fifteen years of Antarctic micrometeorite research by the Italian Programma Nazionale delle Ricerche in Antartide. *EPSC Abstracts* **12**, EPSC2018-1012. https://meetingorganizer.copernicus.org/EPSC2018/EPSC2018-1012.pdf
[22] KBS World (2014) Large Meteoritc Found Near S. Korea's Antarctic Research Station. https://world.kbs.co.kr/service/news_view.htm?lang=e&Seq_Code=107148 (accessed 02/01/2024).
[23] George Delisle, Ian Franchi, Antonio Rossi and Rainer Weiler (1993) Meteorite finds by EUROMET near Frontier Mountain, North Victoria Land Antarctica. *Meteoritics*, **28**, 126–129. https://articles.adsabs.harvard.edu/pdf/1993Metic..28..126D (accessed 02/01/2024).
[24] Expedition to uncover the "Lost" meteorites of Antarctica (2027) University of Manchester. https://www.manchester.ac.uk/discover/news/expedition-uncover-lost-meteorites-antarctica/ (accessed 02/01/2024).
[25] Lunar Meteorite. Allan Hills A81005, Some Meteorite Information. Washington University in St. Louis. https://sites.wustl.edu/meteoritesite/items/lm_alha_a81005/ (accessed 02/01/2024).
[26] Half-Baked Asteroids Have Earth-Like Crust (2009) EurekAlert New Release. https://www.eurekalert.org/news-releases/622520 (accessed 02/01/2024).
[27] Edward R. D. Scott; Richard C. Greenwood; Ian A. Franchi and Ian S. Sanders (2009) Oxygen isotopic constraints on the origin and parent bodies of eucrites, diogenites, and howardites. *Geochimica et Cosmochimica Acta* **73**, 5835–5853. https://doi.org/10.1016/j.gca.2009.06.024
[28] A. H. Treiman (1996) Fossil Life in ALH 84001? Lunar and Planetary Institute. https://www.lpi.usra.edu/lpi/meteorites/life.html (accessed 02/01/2024).
[29] C. Taylor (2022) Can Meteorites on Earth Point to Ancient Life on Mars? Science Friday. https://www.sciencefriday.com/segments/mars-meteorites/ (accessed 02/01/2024).
[30] Every Rock Tells a Story NASA Astromaterials 3D ALH 84001, 55 Martian Meteorite. https://ares.jsc.nasa.gov/astromaterials3d/sample-details.htm?sample=ALH84001-55#:~:text=Formed%20in%20the%20igneous%20plutonic,shock%20from%20multiple%20impact%20events (accessed 02/01/2024).

[31] J. Marchant (2020) Life on Mars: The Story of Meteorite ALH 84001. BBC Science Focus. https://www.sciencefocus.com/space/life-on-mars-the-story-of-meteorite-alh84001/ (accessed 02/01/2024).

CHAPTER 13

[1] The Little Prince's Last Flight: The Story of Antoine de Saint-Exupéry (2020) The National WWII Museum, New Orleans. https://www.nationalww2museum.org/war/articles/the-little-prince-antoine-de-saint-exupery (accessed 02/01/2024).

[2] *7 Timeless Life Lessons from the Little Prince*. Penguin Books. https://www.penguin.co.uk/articles/childrens-article/7-timeless-life-lessons-from-the-little-prince (accessed 02/01/2024).

[3] K. Lohnes and P. Bauer (2023) The Little Prince Fable by Saint-Exupery. Britannica (retrieved July 2023) https://www.britannica.com/topic/The-Little-Prince (accessed 02/01/2024).

[4] M. Buckley (2004) Mysterious Wartime Death of French Novelist. BBC News Channel from Our Own Correspondent. https://news.bbc.co.uk/1/hi/programmes/from_our_own_correspondent/3541662.stm (accessed 02/01/2024).

[5] *Wind, Sand and Stars by Antoine de Saint-Exupéry*. Penguin Books. https://www.penguin.co.uk/books/57080/wind-sand-and-stars-by-antoine-de-saint-exuperytranslated-with-an-introduction-by-william-rees/9780141183190 (accessed 02/01/2024).

[6] C. T. Pillinger (1992) Euromet-a programme for the collection of new extraterrestrial samples: progress, plans, or pipe-dreams? *Meteoritics*, **27**, 277 (abstract). https://adsabs.harvard.edu/full/1992Metic..27..277P

[7] N. Heath and C. Dennis (2023) Beagle 2 Mars Mission Leader Remembered on Launch Anniversary. BBC News Bristol. https://www.bbc.co.uk/news/uk-england-bristol-65788431 (accessed 02/01/2024).

[8] Q. Müller and S. Castelier (2017) Morocco's Stone Rush: Hunting Meteorites Is Big Business for Nomads. Middle East Eye. https://www.middleeasteye.net/features/moroccos-stone-rush-hunting-meteorites-big-business-nomads (accessed 02/01/2024).

[9] Lunar Sample Laboratory Facility. NASA Curation Lunar. https://curator.jsc.nasa.gov/lunar/lun-fac.cfm# (accessed 02/01/2024).

[10] Sample Requests for Research. NASA Curation Lunar. https://curator.jsc.nasa.gov/lunar/sampreq/requests.cfm (accessed 02/01/2024).

[11] J. Maslin (13 July 2011) Supposition as Research: A Sort-of-True Story about NASA and a Thief. The New York Times. https://www.nytimes.com/2011/07/14/books/sex-on-the-moon-by-ben-mezrich-book-review.html (accessed 02/01/2024).

[12] R. Korotev. Some Meteorite Information Washington University in St Louis. Lunar Meteorite: Yamato 791197. https://sites.wustl.edu/meteoritesite/items/lm_yamato_791197/ (accessed 02/01/2024).

[13] R. Korotev. Some Meteorite Information Washington University in St Louis. Lunar Meteorite: Allan Hills A81005. https://sites.wustl.edu/meteoritesite/items/lm_alha_a81005/ (accessed 02/01/2024).

[14] U. B. Marvin (1983) The discovery and initial characterization of Allan Hills A81005: the first lunar meteorite. *Geophysical Research Letters*, **10**, 775–778. https://agupubs.onlinelibrary.wiley.com/doi/abs/10.1029/GL010i009p00775

[15] K. Yanai and H. Kojima (1984) Yamato-791197: A lunar meteorite in the Japanese collection of Antarctic meteorites. *Memoirs of the National Institute of Polar Research*, Special Issue **35**, 18–34. https://nipr.repo.nii.ac.jp/?action=repository_action_common_download&item_id=1716&item_no=1&attribute_id=18&file_no=1

[16] Meteoritical Bulletin Database. https://www.lpi.usra.edu/meteor/metbull.php (accessed 02/01/2024).
[17] Christie's Deep Impact Lunar and Rare Meteorites 12–25 August 2020. https://www.christies.com/about-us/press-archive/details?PressReleaseID=9742&lid=1 (accessed 02/01/2024).
[18] Christie's Meteorites from the Collection of Michael Farmer (23 March to 6 April 2022) Online 21125. https://onlineonly.christies.com/s/meteorites-collection-michael-farmer/lots/2182 (accessed 02/01/2024).
[19] Sotheby's Meteorites (27 July 2022) Select Specimens from the Moon, Mars, Vesta and More. New York. https://www.sothebys.com/en/buy/auction/2022/meteorites-select-specimens-from-the-moon-mars-vesta-and-more (accessed 02/01/2024).
[20] Bonhams (18–28 May 2021) Meteorites, Tektites and Impact Memorabilia. https://www.bonhams.com/auction/27190/meteorites-tektites-and-impact-memorabilia/ (accessed 02/01/2024).
[21] J. Ardis (2020) Meteorites at Auction: A Category Spotlight Auction Daily. https://auctiondaily.com/news/meteorites-at-auction-a-category-spotlight/ (accessed 02/01/2024).
[22] ebay. Collectable Meteorites & Tektites. https://www.ebay.co.uk/b/Collectable-Meteorites-Tektites/3239/bn_2316367 (accessed 02/01/2024).
[23] Tucson Gem and Mineral Show. The Largest, Oldest and Most Prestigious Gem and Mineral Show in the World. https://www.tgms.org/show (accessed 02/01/2024).
[24] Ensisheim Meteorite Show. https://www.facebook.com/EnsisheimMeteoriteShow/ (accessed 02/01/2024).
[25] R. Korotev. Lunar Meteorites. Everything You Need to Know Washington University in St Louis. https://sites.wustl.edu/meteoritesite/items/lunar-meteorites/#:~:text=Any%20rock%20on%20the%20lunar,into%20orbit%20around%20those%20bodies (accessed 02/01/2024).
[26] NWA 7034 (Black Beauty) Martian Meteorites Collection. The Virtual Microscope. https://www.virtualmicroscope.org/content/nwa-7034-black-beauty (accessed 02/01/2024).
[27] J. Lipton (2017) These Guys Hunt for Space Rocks, and Sell Them for Enormous Profit to Collectors. CNBC. https://www.cnbc.com/2017/01/03/these-guys-hunt-for-space-rocks-and-sell-them-for-enormous-profit-to-collectors.html (accessed 02/01/2024).

CHAPTER 14

[1] Deep Impact (1998) American Science Fiction Disaster Movie. https://en.wikipedia.org/wiki/Deep_Impact_(film) (accessed 02/01/2024).
[2] Armageddon (1998) American Science Fiction Disaster Movie. https://en.wikipedia.org/wiki/Armageddon_(1998_film) (accessed 02/01/2024).
[3] C. M. O'D. Alexander (2017) The origin of inner Solar System water. *Philosophical Transactions of the Royal Society*, **375**, 20150384. https://doi.org/10.1098/rsta.2015.0384
[4] A. Piani, Y. Marrocchi, T. Rigaudier, L. Vacher, D. Thomassin and B. Marty (2020) Earth's water may have been inherited from material similar to enstatite chondrite meteorites. *Science*, **369**, 1110–1113. https://www.science.org/doi/10.1126/science.aba1948
[5] Space Weathering on Airless Bodies. NASA Solar System Exploration Research Virtual Institute. https://sservi.nasa.gov/articles/space-weathering-on-airless-bodies/ (accessed 02/01/2024).
[6] T. H. Burbine, T. J. McCoy, E. Jarosewich and J. M. Sunshine (2003) Deriving asteroid mineralogies from reflectance spectra: implications for the MUSES-C target asteroid. *Antarctic Meteorite Research*, **16**, 185–195. National Institute of Polar Research. https://core.ac.uk/download/pdf/51485481.pdf

[7] Lunar Rocks and Soils from Apollo Missions. NASA Lunar Curation. https://curator.jsc.nasa.gov/lunar/#:~:text=Between%201969%20and%201972%20six,exploration%20sites%20on%20the%20Moon (accessed 02/01/2024).
[8] J. Mehta (2018) How the Apollo Missions Transformed Our Understanding of the Moon's Origin. The Planetary Society. https://www.planetary.org/articles/apollo-moon-origin (accessed 02/01/2024).
[9] Aron Gronstal. Re-thinking a Critical Period in Earth's History. Astrobiology at NASA. https://astrobiology.nasa.gov/news/re-thinking-a-critical-period-in-earths-history/ (accessed 02/01/2024).
[10] O. J. Tucker, W. M. Farrell, R. M. Killen and D. M. Hurley (2019) Solar wind implantation into the lunar regolith: Monte Carlo simulations of H retention in a surface with defects and the H2 exosphere. *Journal of Geophysical Research: Planets*, **124**, 278–293. https://doi.org/10.1029/2018JE005805
[11] Jonathan Amos (2020) China's Chang'e-5 Mission Returns Moon Samples. BBC News. https://www.bbc.co.uk/news/science-environment-55323176 (accessed 02/01/2024).
[12] Comets: Facts. NASA Solar System Exploration. https://solarsystem.nasa.gov/asteroids-comets-and-meteors/comets/in-depth/ (accessed 02/01/2024).
[13] Kuiper Belt: Facts. NASA Solar System Exploration. https://solarsystem.nasa.gov/solar-system/kuiper-belt/in-depth/ (accessed 02/01/2024).
[14] Oort Cloud: Facts. NASA Solar System Exploration. https://solarsystem.nasa.gov/solar-system/oort-cloud/in-depth/ (accessed 02/01/2024).
[15] Stardust NASA's Comet Mission. https://solarsystem.nasa.gov/stardust/tech/aerogel.html (accessed 02/01/2024).
[16] NASA Stardust/Stardust NExT Missions. https://science.nasa.gov/mission/stardust/ (accessed 02/01/2024). Stardust (Spacecraft). Wikipedia. https://en.wikipedia.org/wiki/Stardust_(spacecraft) (accessed 02/01/2024).
[17] The Solar Wind across Our Solar System. NASA Solar System Exploration. https://solarsystem.nasa.gov/resources/2288/the-solar-wind-across-our-solar-system/ (accessed 02/01/2024).
[18] Genesis Search for Origins. NASA JPL. https://solarsystem.nasa.gov/genesismission/gm2/news/features/wrapup.htm (accessed 02/01/2024).
[19] Kerry Ellis (2006) Genesis: Learning from Mistakes. NASA Appel. https://appel.nasa.gov/2006/04/01/genesis-learning-from-mistakes/ (accessed 02/01/2024).
[20] T. Malik (2004) NASA Researchers, Stunt Pilots Prepare for Genesis Probe's Return. https://www.space.com/271-nasa-researchers-stunt-pilots-prepare-genesis-probe-return.html (accessed 02/01/2024).
[21] S. Crowther (2012) A Jigsaw Puzzle from Space. Earth and Solar System. https://earthandsolarsystem.wordpress.com/2012/07/10/a-jigsaw-puzzle-from-space/ (accessed 02/01/2024).
[22] S. Crowther (2012) NASA's Genesis Mission: Oxygen Earth and Solar System. https://earthandsolarsystem.wordpress.com/2013/04/25/nasas-genesis-mission-oxygen/ (accessed 02/01/2024).
[23] R. N. Clayton (2002) Self-shielding in the solar nebula. *Nature*, **415**, 860–861. https://www.nature.com/articles/415860b
[24] Hayabusa. Wikipedia. https://en.wikipedia.org/wiki/Hayabusa (accessed 02/01/2024).
[25] Hayabusa: Mission Accomplished (2010) Space News. https://spacenews.com/hayabusa-mission-accomplished/ (accessed 02/01/2024).
[26] BBC News website (2010) Hayabusa Capsule Particles May Be from Asteroid. https://www.bbc.co.uk/news/10519895 (accessed 02/01/2024).
[27] T. Yada et al. (2013) Hayabusa-returned sample curation in the planetary material sample curation facility of JAXA. *Meteoritics and Planetary Science*, **49**, 135–153. https://doi.org/10.1111/maps.12027

[28] T. Nakamura et al. (2011) Itokawa dust particles: a direct link between S-type asteroids and ordinary chondrites. *Science*, **333**, 1113–1116. https://www.science.org/doi/10.1126/science.1207758
[29] Asteroid Explorer Hayabusa2. JAXA. https://www.isas.jaxa.jp/en/missions/spacecraft/current/hayabusa2.html (accessed 02/01/2024).
[30] K. Kitazato et al. (2019) The surface composition of asteroid 162173 Ryugu from Hayabusa2 near-infrared spectroscopy. *Science*, **364**, 272–275. https://www.science.org/doi/10.1126/science.aav7432
[31] T. Nakmura (2005) Post-hydration thermal metamorphism of carbonaceous chondrites. *Journal of Mineralogical Petrological Science*, **100**, 260–272. https://www.jstage.jst.go.jp/article/jmps/100/6/100_6_260/_pdf/-char/en
[32] A. J. King et al. (2019) The Yamato-type (CY) carbonaceous chondrite group: analogues for the surface of asteroid Ryugu? *Chemie der Erde – Geochemistry*, **79**(4), 125531. https://doi.org/10.1016/j.chemer.2019.08.003
[33] Hayabusa2, Japan's Mission to Ryugu and Other Asteroids. The Planetary Society. https://www.planetary.org/space-missions/hayabusa2 (accessed 02/01/2024).
[34] P. Rincon (2020) Hayabusa-2: Pieces of an Asteroid Found inside Space Capsule. BBC News. https://www.bbc.co.uk/news/science-environment-55315502 (accessed 02/01/2024).
[35] Toru Yada et al. (2021) Preliminary analysis of the Hayabusa2 samples returned from C-type asteroid Ryugu. *Nature Astronomy*, **6**, 214–220. https://doi.org/10.1038/s41550-021-01550-6
[36] Motoo Ito et al. (2022) A pristine record of outer Solar System materials from asteroid Ryugu's returned sample. *Nature Astronomy*, **6**, 1163–1171. https://doi.org/10.1038/s41550-022-01745-5
[37] Jean-Alix Barrat, B. Zanda, Frederic Moynier, Claire Bollinger, C. Liorzou, Germain Bayon (2012) Geochemistry of CI chondrites: major and trace elements, and Cu and Zn isotopes. *Geochimica et Cosmochimica Acta*, **83**, 79–92. https://core.ac.uk/download/pdf/52731265.pdf
[38] Katharina Lodders (2003) Solar System abundances and condensation temperatures of the elements. *The Astrophysical Journal*, **591**, 1220–1247. https://iopscience.iop.org/article/10.1086/375492
[39] Christopher D. K. Herd (2023) Analyzing asteroid Ryugu. *Science*, 379, 784–785. https://www.science.org/doi/abs/10.1126/science.ade4188
[40] Richard C. Greenwood et al. (2022) Oxygen isotope evidence from Ryugu samples for early water delivery to Earth by CI chondrites. *Nature Astronomy*, **7**, 29–38. https://doi.org/10.1038/s41550-022-01824-7
[41] OSIRIS-Rex. NASA Solar System Exploration. https://solarsystem.nasa.gov/missions/osiris-rex/in-depth/
[42] Kevin J. Walsh (2018) Rubble pile asteroids. *Annual Review of Astronomy and Astrophysics* **56**, 593–624. https://ui.adsabs.harvard.edu/abs/2018ARA%26A..56..593W/abstract
[43] OSIRIS-REx Mission NASA In depth. https://science.nasa.gov/mission/osiris-rex/in-depth/ (accessed 02/04/2024).
[44] Jonathan Amos (2020) Elation as Nasa's Osiris-Rex Probe Tags Asteroid Bennu in Sample Bid. BBC News. https://www.bbc.co.uk/news/science-environment-54624653 (accessed 02/04/2024).
[45] NASA Explore (2020) OSIRIS-REx TAGs Surface of Asteroid Bennu. https://www.nasa.gov/solar-system/osiris-rex-tags-surface-of-asteroid-bennu/ (accessed 02/04/2024).
[46] Jonathan Amos (2023) Osiris-Rex: Nasa Confirms Return of Asteroid Bennu Samples. BBC News. https://www.bbc.co.uk/news/science-environment-66893661 (accessed 02/01/2024).

[47] Bringing Mars Samples to Earth. Mars Sample Return Mission. NASA Science. https://mars.nasa.gov/msr/ (accessed 02/01/2024).
[48] Phobos in Depth. NASA Solar System Exploration. https://solarsystem.nasa.gov/moons/mars-moons/phobos/in-depth/ (accessed 02/01/2024).
[49] JAXA MMX. Martian Moons Exploration. https://www.mmx.jaxa.jp/en/ (accessed 02/01/2024).
[50] N. T. Tillman (2014) Potato-Shaped Mars Moon Phobos May Be a Captured Asteroid. https://www.space.com/24285-mars-moon-phobos-captured-asteroid.html (accessed 02/01/2024).
[51] P. Rosenblatt et al. (2016) Accretion of phobos and deimos in an extended debris disc stirred by transient moons. *Nature Geoscience*, **9**, 581–583. https://doi.org/10.1038/ngeo2742
[52] Phobos-Grunt. NASA Solar System Exploration. https://solarsystem.nasa.gov/missions/phobos-grunt/in-depth/ (accessed 02/01/2024).
[53] NASA and ESA Agree on Next Steps to Return Mars Samples to Earth. NASA Jet Propulsion Laboratory. https://www.jpl.nasa.gov/news/nasa-and-esa-agree-on-next-steps-to-return-mars-samples-to-earth (accessed 02/01/2024).

CHAPTER 15

[1] H. Y. McSween Jr. (1994) What we have learned about Mars from the SNC meteorites. *Meteoritics*, **29**, 757–779. https://articles.adsabs.harvard.edu/pdf/1994Metic..29..757M
[2] Z. Váci and C. Agee (2020) Constraints on Martian chronology from meteorites. Geosciences, **10**, 455. https://www.mdpi.com/2076-3263/10/11/455
[3] In recent years, a number of achondrites (meteorites that lack chondrules) have been identified with very old ages. Like the meteorites from Mars, they formed from molten rock that was produced on their source bodies by melting of precursor material. But unlike Mars, this melting event took place very shortly after the formation of the Solar System. Their source bodies would have been asteroids, which are much smaller than Mars, and so not capable of sustaining long-term volcanic activity. A good example of one of these old achondrites is Erg Chech 002 which was found in the Sahara Desert in 2020. It has been dated as having formed less than two million years after the formation of the Solar System, which as we saw in Chapter 11, is dated at 4,567 million years ago; L. Crane (2021) 4.6-Billion-Year-Old Meteorite Is the Oldest Volcanic Rock Ever Found. New Scientist. https://www.newscientist.com/article/2270314-4-6-billion-year-old-meteorite-is-the-oldest-volcanic-rock-ever-found/ (accessed 02/01/2024); Meteoritical Bulletin Database Entry for Erg Chech 002. https://www.lpi.usra.edu/meteor/metbull.php?code=72475 (accessed 02/01/2024).
[4] D. D. Bogard and P. Johnson (1983) Martian gases in an Antarctic meteorite. *Science*, **221**, 651–654. https://www.science.org/doi/10.1126/science.221.4611.651
[5] P. G. Conrad et al. (2016) In situ measurement of atmospheric krypton and xenon on Mars with Mars Science Laboratory. *Earth and Planetary Science Letters*, **454**, 1–9. https://www.sciencedirect.com/science/article/pii/S0012821X16304514
[6] A. H. Treiman, J. D. Gleason and D. D. Bogard (2000) The SNC meteorites are from Mars. *Planetary and Space Science*, **48**, 1213–1230. https://www.sciencedirect.com/science/article/pii/S0032063300001057
[7] R. N. Clayton and T. K. Mayeda (1983) Oxygen isotopes in eucrites, shergottites, nakhlites, and chassignites. *Earth Planetary Science Letters*, **62**, 1–6. https://www.sciencedirect.com/science/article/abs/pii/0012821X83900663

[8] As we looked at in Chapter 10, oxygen has three stable isotopes: oxygen-16 (^{16}O), oxygen-17 (^{17}O) and oxygen-18 (^{18}O). Compared to the Earth, Moon, or Vesta rocks from Mars have a slightly lower amount of ^{16}O. This turns out to be a useful characteristic to determine whether a sample is Martian or not. Because planets melt, they tend to get totally mixed up due to this early large scale melting and any natural variation in oxygen isotopes gets averaged out. So, Martian rocks have a very specific isotopic value that can be used as a way of fingerprinting them. This was fine until the NWA 7034, otherwise known as Black Beauty, turned up. See main text for further discussion.

[9] L. C. Bouvier et al. (2018) Evidence for extremely rapid magma ocean crystallization and crust formation on Mars. *Nature*, **558**, 586–589. https://doi.org/10.1038/s41586-018-0222-z

[10] C. B. Agee et al. (2013) Unique meteorite from early Amazonian Mars: water-rich basaltic breccia Northwest Africa 7034. *Science*, **339**, 780–785. https://www.science.org/doi/abs/10.1126/science.1228858

[11] A. Goodwin, R. J. Garwood and R. Tartese (2022) A review of the "Black Beauty" Martian Regolith Breccia and its Martian habitability record. *Astrobiology*, **22**. https://www.liebertpub.com/doi/pdf/10.1089/ast.2021.0069

[12] J. A. Cartwright, U. Ott, S. Herrmann and C. B. Agee (2014) Modern atmospheric signatures in 4.4 Ga Martian meteorite NWA 7034. *Earth and Planetary Science Letters*, **400**, 77–87. https://www.sciencedirect.com/science/article/pii/S0012821X14003021

[13] J. Farquhar, M. H. Thiemens and T. Jackson (1998) Atmosphere-surface interactions on mars: Δ17O measurements of carbonate from ALH 84001. *Science*, **280**, 1580–1582. https://www.science.org/doi/10.1126/science.280.5369.1580

[14] NASA's Perseverance Rover Deposits First Sample on Mars Surface. NASA Science Mars Exploration. https://mars.nasa.gov/news/9323/nasas-perseverance-rover-deposits-first-sample-on-mars-surface/ (accessed 02/01/2024).

[15] Perseverance's "Three Forks" Sample Depot Map. NASA Science Mars Exploration. https://mars.nasa.gov/news/9323/nasas-perseverance-rover-deposits-first-sample-on-mars-surface/ (accessed 02/01/2024).

[16] W. S. Cassata et al. (2018) Chronology of Martian breccia NWA 7034 and the formation of the Martian crustal dichotomy. *Sciences Advances*, **4**, eaap8306. https://www.science.org/doi/pdf/10.1126/sciadv.aap8306

CHAPTER 16

[1] E. T. Dunham et al. (2023) Calcium-aluminum-rich inclusions in non-carbonaceous chondrites: abundances, sizes, and mineralogy. *Meteoritics and Planetary Science*, **58**, 643–671. https://ui.adsabs.harvard.edu/abs/2023M%26PS...58..643D/abstract

[2] A. P. Boss, C. M. O'D. Alexander and M. Podolak (2020) Evolution of CAI-sized particles during FU Orionis outbursts. I. Particle trajectories in protoplanetary disks with beta cooling. *The Astrophysical Journal*, **901**, 81. https://iopscience.iop.org/article/10.3847/1538-4357/abafb9/pdf

[3] L. Grossman (1972) Condensation in the primitive solar nebula. *Geochimica et Cosmochimica Acta*, **36**, 597–619. https://www.sciencedirect.com/science/article/pii/0016703772900786

[4] G. Arrhenius and B. R. De (1973) Equilibrium condensation in a solar nebula. *Meteoritics*, **8**, 297–313. https://articles.adsabs.harvard.edu/pdf/1973Metic...8..297A

[5] L. Grossman and J. W. Larimer (1974) Early chemical history of the solar system. *Reviews in Geophysics*, **12**, 71–101. https://doi.org/10.1029/RG012i001p00071

[6] E. Anders and E. Zinner (1993) Interstellar grains in primitive meteorites: diamond, silicon carbide, and graphite. *Meteoritics*, **28**, 490–514. https://articles.adsabs.harvard.edu/pdf/1993Metic..28..490A

[7] S. Amari (2014) Recent progress in presolar grain studies. *Mass Spectrometry*, **5**, S0042. https://www.jstage.jst.go.jp/article/massspectrometry/3/Special_Issue_3/3_S0042/_html/-char/en

[8] D. D. Clayton (1979) Supernovae and the origin of the Solar System Space. *Science Reviews*, **24**, 147–226. https://articles.adsabs.harvard.edu/pdf/1979SSRv...24..147C

[9] Presolar Grains. Wikipedia. https://en.wikipedia.org/wiki/Presolar_grains (accessed 02/01/2024).

[10] T. Ming et al. (1988) Isotopic anomalies of Ne, Xe, and C in meteorites. I. Separation of carriers by density and chemical resistance. *Geochimica et Cosmochimica Acta*, **24**, 147–226. https://www.sciencedirect.com/science/article/pii/0016703788902761?via%3Dihub

[11] A. M. Davis (2011) Stardust in meteorites. PNAS, **108**, 19142–19146. https://www.pnas.org/doi/full/10.1073/pnas.1013483108

[12] P. Heck et al. (2020) Lifetimes of interstellar dust from cosmic ray exposure ages of presolar silicon carbide. *PNAS*, **117**, 1884–1889. file:///C:/Users/Richard/Downloads/pnas.1904573117.pdf

[13] P. Mueller and J. Vervoort. Secondary Ion Mass Spectrometer (SIMS). https://serc.carleton.edu/msu_nanotech/methods/SIMS.html#:~:text=As%20a%20class%2C%20SIMS%20instruments,and%20are%20referred%20to%20as (accessed 02/01/2024).

[14] K. D. McKeegan (2007) Ernst Zinner, lithic astronomer. *Meteoritics and Planetary Science*, **42**, 1045–1054. https://adsabs.harvard.edu/pdf/2007M&PS...42.1045M

[15] E. Zinner (2014) Presolar Grains. In: *Treatise on Geochemistry 2nd Edition*. https://presolar.physics.wustl.edu/Laboratory_for_Space_Sciences/Publications_2014_files/Zinner13c.pdf

[16] A supernova is essentially a giant cosmic explosion that takes place when certain types of stars reach the end of their lives. Two main types of supernova are recognised by astronomers. The first type involves binary star systems. Unlike our Sun, the majority of stars are in fact binary systems consisting of two stars orbiting around a single point. Where one of the two stars is a white dwarf, under certain circumstances, it can grab material from its partner star, become unstable and explode. A white dwarf is a dense remnant of a Sun mass star that has exhausted all its fuel. Astronomers call this variety of supernova a Type I. The second type of supernova involves stars that are significantly more massive than our Sun. When all their nuclear fuel has been used up their cores collapse. This results in a giant explosion. It is during this event that presolar diamond and graphite are produced. This type of supernova is classified as a Type II. It is also known as a core-collapse supernova. For further information on supernovae visit: Imagine the Universe, NASA. https://imagine.gsfc.nasa.gov/science/objects/supernovae2.html (Accessed 02/01/2024)

[17] C. Ruth (2009) Where Do Chemical Elements Come from? ChemMatters. https://www.frontiercsd.org/cms/lib/NY19000265/Centricity/Domain/147/chemmattersWhere%20Elements-oct2009.pdf (accessed 02/01/2024).

Index

Note: *Italic* page numbers refer to figures.